Fluidized-Bed Energy Technology:

Coming to a Boil

Fluidized-Bed Energy Technology:

Coming to a Boil

by Walter C. Patterson
and Richard Griffin

an INFORM Report

INFORM
25 Broad Street
New York, N.Y. 10004
212/425-3550

ISBN 0-918780-10-1

Library of Congress #78-60484

*Cover photo courtesy of Babcock Contractors Inc./
Combustion Systems Limited*

This report was prepared in part under contract with the United States
Government. Neither the United States nor the United States Department
of Energy, nor any of their employees, nor any of their contractors,
subcontractors, or their employees, makes any warranty, express or
implied, or assumes any legal liability or responsibility for the accuracy
or completeness of any information, apparatus, product or process
disclosed or represents that its use would not infringe on privately owned
rights. The views and conclusions contained in this report should not be
interpreted as representing the official policies, either expressed or
implied, of the United States Department of Energy.

INFORM is a nonprofit, tax-exempt organization, established in 1973,
which conducts research on the impact of American corporations on the
environment, employees, and consumers. INFORM publishes books, condens-
ed reports, and newsletters. These seek to clarify and define the nature of
some of today's most serious corporate social problems. They describe and
evaluate programs and practices that industries could adopt to improve future
social performance. INFORM's program is supported by subscriptions and
contributions from foundations, corporations, financial institutions, universities, government agencies, and concerned individuals.

Contents

Preface

The dwindling supplies of oil and natural gas have resulted in plans to shift to alternate fuels. The expanded use of coal holds promise for bringing about this needed change in direction.

Fluidized-bed combustion may play a significant role in permitting the burning of coal in an environmentally acceptable manner, a step forward towards the widespread use of clean technologies. Additionally, this technology offers the user flexibility in burning virtually any type of fuel, individually or simultaneously, without significant boiler modifications.

This report reviews the worldwide status of research, development, demonstration, and commercialization of fluidized-bed combustion. It provides usable, working knowledge for designers, users and educators. I hope that this exchange of knowledge serves to stimulate new ideas which will result in extending application of this technology.

John F. O'Leary
Deputy Secretary
U.S. Department of Energy

Acknowledgments

The authors would like to express their gratitude to those people, without whose help this study would not have been possible. We would especially like to thank INFORM's Executive Director, Joanna Underwood, for her seemingly endless supply of energy and inspiration; and Jean Halloran, INFORM's Editorial Director, whose ability to clarify the most complex material is one of the study's greatest assets.

Several experts in fluidized-bed energy technology reviewed a preliminary draft of the manuscript. We would like to acknowledge the insightful comments of Dr. Steven I. Freedman, John Smithson and Sam Biondo of the U.S. Department of Energy; Morris Altschuler, U.S. Environmental Protection Agency; Shelton Ehrlich, Electric Power Research Institute; David Comey, Citizens for a Better Environment; Gerald Decker, Dow Chemical Company; Amory Lovins, Friends of the Earth; and Arthur Squires, Virginia Polytechnic and State University. We would also like to thank two members of INFORM's Board--Howard Hardesty and Dennis Meadows--for their help and advice throughout the project.

We appreciate the financial support given this study by the U.S. Department of Energy and the Rockefeller Brothers Fund. Total responsibility for the final manuscript rests, of course, with INFORM.

Special thanks to Virginia Jones, INFORM's Administrator, who kept the office running even during New York's worst snow storm; Deborah Moses, for careful and precise editing and proofreading; Susan Jakoplic, for appendix, glossary and production aid; Alex Reese and Peter Rossbach for preliminary research on the project; Greg Cohen, whose expertise with corporations, charm and wit on the telephone helped us immensely; Daniel Wiener, for aid rendered in proofreading and production; and talented typists Mary Ferguson, John Klingberg, Bill Funk, and Albert Dlugash for their wonderful patience with copy. Finally, to the staff of INFORM, a thank you for making the office such a fine place to work.

Part 1

Introduction/Findings

Introduction/Findings

Overview

Some new problems may prove to have old solutions. When world oil prices leapt up and natural gas supplies became unreliable, energy users at first sought novel alternative sources: oil shales, tar sands, solar energy, wind energy, geothermal energy, nuclear energy and fast breeder reactors. Many of these supply technologies do indeed appear promising. However, some bring with them heavy environmental costs, and most are years away from wide commercial availability. More recently it has become clear that coal -- once the world's predominant fuel, but overlooked in the early months of the so-called "energy crisis" -- will have a key role to play in the transition from dwindling to sustainable sources of energy.

Estimates of world fossil fuel reserves are notoriously variable. There is nevertheless agreement that reserves of coal are orders of magnitude larger than reserves of petroleum or natural gas. Coal is, to be sure, a troublesome material to obtain and use. Wherever it is mined, its extraction creates problems of worker health and safety and of environmental degradation. Compared to oil or gas, it is an awkward material to transport and store. If it is burned, it produces not only carbon dioxide but also sulfur oxides, nitrogen oxides, particulates and other air pollutants. It does not burn completely, but leaves behind substantial quantities of solid waste in the form of ash. Indeed, these negative characteristics contributed to its fall from grace in favor of oil and gas.

It is therefore hardly surprising that energy users have been unenthusiastic about switching back to coal. However, the rise in oil and gas prices, the increase in oil importation, and growing problems with nuclear power have brought about a renaissance in coal research and development, seeking

3

improvements in coal extraction and utilization. Some of the longer-term possibilities for advanced coal utilization, including coal gasification and liquefaction, were described in ENERGY FUTURES: Industry and the New Technologies, by Stewart Herman and James Cannon (INFORM/Ballinger 1977). One coal technology, however, has attracted an upsurge of immediate interest. It is called "fluidized-bed combustion." Fluidized-bed combustion is a versatile technical innovation which makes it possible to burn coal -- and also other troublesome fuels -- more cleanly, and to produce heat, steam and electricity more efficiently than conventional coal-burning systems at a reasonable cost.

Although its history is much shorter than that of coal usage, fluidized-bed technology for non-energy applications has been on the scene for many decades. However, its use as an energy technology is of much more recent vintage; indeed, on a commercial scale, this use is still in the future. But many now believe this future to be overdue and arriving rapidly. The last two years have seen a dramatic rise in government and corporate spending on fluidized-bed energy technology. Those involved are now moving on from research and development to large-scale investment.

Very few people outside the specialist field, however, have even heard of fluidized-bed combustion. The present report is an attempt to rectify that omission: to give interested readers an introduction to fluidized-bed energy technologies, especially fluidized-bed combustion. The following sections will describe the basic principles of fluidized-bed applications envisioned, the economic, environmental and institutional implications, and the present prognosis for future development. There then follows a series of nineteen profiles of organizations involved in fluidized-bed energy technologies.

INFORM's survey of companies active in this field suggests that fluidized-bed combustion is on the threshold of becoming a major energy technology. Several firms in the United States, Norway, and the United Kingdom are now commercially offering fluidized-bed units for small-scale industrial applications (some companies are offering units up to 500,000 pounds per hour), and a number have been installed and are operating. The firms offering the systems are: the Foster Wheeler Corporation, Energy Resources Company, Fluidyne Engineering Corporation (all United States); Stone-Platt Fluidfire Limited, Babcock & Wilcox, Limited, Energy Equipment Company Limited (all United Kingdom); and Mustad and Søn (Norway) (see profiles and appendix for details). Research is continuing in the United States, United Kingdom, Sweden, Norway, West Germany, Japan, India, the Netherlands, and a number of other countries, both by government and by industry. The technology is being developed for applications ranging from large utility boilers to systems small enough for domestic use. Those involved in development include boiler manufacturers like Babcock & Wilcox, Limited, Foster Wheeler Corporation, Combustion Engineering, Inc., and the Babcock & Wilcox Company, turbine manufacturers like Stal-Laval Turbin AB and Curtiss-Wright Corporation, electrical engineering manufacturers like the General Electric Company and Bharat Heavy Electricals, Ltd., research and development organizations like the British National Coal Board Coal Research Establishment, Bergbau-Forschung of West Germany, and the Oak Ridge and Argonne National Laboratories in the United States, as well as government departments in the United States, Britain, West Germany and elsewhere.

Test fluidized-bed systems ranging from laboratory-size models to small power-generating stations are operating in dozens of locations in the United States, Britain and elsewhere. Seven large-scale industrial prototypes are under construction in the U.S. and Britain. Commercial orders have been placed for adapting fluidized-bed systems to existing boiler plants; some of these retrofits are already operating, while others will come on stream in 1978 and 1979. Several major electrical utilities, including American Electric Power (see profile) and British Columbia Hydro, have commissioned feasibility and design studies for full-scale plants which are in various stages of consideration; orders for utility demonstration plants may be placed in 1979.

Fluidized-bed combustion systems appear to promise a simple, elegant, and versatile approach to burning virtually any combustible material. Their advantages in controlling sulfur emissions are especially noteworthy. Because of the distinctive behavior of a fluidized bed, a fluidized-bed combustion unit burning high-sulfur coal can demonstrably trap enough of the sulfur to comply with air-quality standards without expensive and potentially unreliable stack-gas cleaning equipment. Other air pollutants like nitrogen oxides can likewise be readily minimized. Sulfur-trapping will involve production of solid waste, but the extra waste is likely to be much more manageable than the wet sludge produced by conventional coal-fired systems equipped with scrubbers.

Some technical problems, relating for instance to the feeding of coal, still appear to trouble United States developers of large-sized fluidized-bed systems; but their European colleagues surveyed by INFORM have found no such difficulties. Indeed, in general, the European experience with development of fluidized-bed combustion seems to be much more straightforward and encouraging than has been the case thus far in the United States.

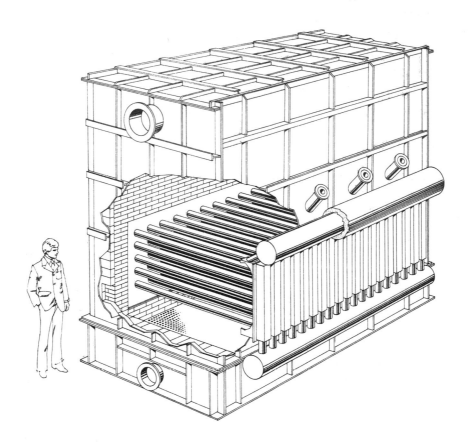

Artist's rendering of industrial demonstration atmospheric fluidized-bed combustor designed by the Fluidyne Engineering Corporation for the Owatowna Tool Company. Drawing: Courtesy Fluidyne Engineering Corporation.

INFORM's survey suggests that decision-makers in the United States would be well advised to consider European evidence as well as American when evaluating prospects for fluidized-bed systems.

Cost estimates for fluidized-bed combustion units are notoriously unreliable and site-specific. In addition, they vary considerably from company to company, not the least according to the size of the system. Any cost figures are founded on conjecture, since there are not enough commercially operating fluidized-bed units to provide data for credible estimates. However, studies completed by the organizations surveyed by INFORM indicate that a fluidized-bed unit will cost the same as or less than an equivalent-size conventional coal-fired system equipped with a scrubber (see Burns and Roe, Inc. and Combustion Engineering, Inc. profiles for examples).

The companies involved in the field suggest that the acceptance of fluidized-bed technology might be hastened in a number of ways. The United States government might reconsider some of its criteria for financial assistance for pollution-control measures, extending them to cover fluidized-bed systems. Such systems cannot presently be financed with low-interest bonds. The United States government could also place orders for appropriate fluidized-bed systems for its own administrative and industrial facilities, to help to develop the market, as could the government of the United Kingdom. As economic and institutional uncertainties are resolved, fluidized-bed combustion appears likely to play a prominent role in energy supply.

The Fluidized Bed: Basic Principles

Imagine a child's playground sandbox, with a floor made of strong porous material. Under the floor of the sandbox is an airtight chamber. Air is pumped under pressure into this chamber. The air emerges upward through the porous floor, and bubbles up through the sand. If the air is rising rapidly enough, it begins to lift the grains of sand, suspending them in a churning turbulent mass which behaves just like a basinful of boiling water. This turbulent mass of solid particles is called a "fluidized bed."

A fluidized bed combines the physical characteristics of both a solid and a liquid. The combination has found many uses in the chemical industry and elsewhere. The application which is of particular interest in this report is the use of the fluidized bed as a medium in which to burn fuel: fluidized-bed combustion. In the firebox of a conventional boiler, fuel burns on a grate, or in mid-air in a cloud of flame. In a "fluidized-bed combustor," however, the bottom of the firebox is filled with granular inert particles of sand, limestone or ash. Air is blown up through holes or pores in the floor of the firebox, making the particles into a fluidized bed. Fuel -- for instance crushed coal -- is fed into this bed. The fuel may constitute less than one percent of the material in the bed. But, as the fuel burns, it makes all the inert particles red hot. The surface of the bed looks like bubbling molten lava. The turbulence of the churning bed keeps the temperature stable, so that the bed does not get rapidly hotter or cooler. Heat is transferred within the bed, and from the bed to the surrounding walls or boiler tubes, by the direct impact of the glowing particles. This direct impact allows a much higher rate of heat transfer than there is in a conventional boiler.

As fresh fuel is added, even though it may be much cooler initially, its temperature rises rapidly to that of the whole hot bed. Accordingly, even very low-quality fuel can be burned: low-grade coal, urban refuse, even wet sludge -- materials which could not be burned in any conventional firebox.

The fluidized bed delivers the heat from the burning fuel to the boiler wall or boiler tubes, which in turn deliver it to the water or steam being heated. The hot water or steam may be used for industrial purposes, to generate electricity, or to heat buildings. Because the heat is delivered by the direct impact of particles, good-quality high-temperature high-pressure steam can be produced while operating a fluidized-bed combustor at a temperature much lower than that in the firebox of a conventional boiler. The low operating temperature makes it possible to reduce dramatically the formation of nitrogen oxides. Furthermore, at such a temperature coal ash, another product of burning coal, does not melt. This avoids the problems caused by molten ash, including corrosion of boiler tubes.

The fluidized bed transfers the heat out of the bed comparatively rapidly. It is therefore possible to produce more heat output per unit time in a given volume of firebox. Accordingly, a fluidized-bed combustor can be physically much smaller than a conventional boiler of the same heat output. This in turn means that the fluidized-bed boiler is likely to cost less, may be built more rapidly, and transported more easily.

If the fuel to be burned in the fluidized-bed combustor is coal, it may contain small but troublesome amounts of sulfur. Such coal, when burned in a conventional boiler, releases sulfur dioxide and sulfur trioxide, noxious gases which have been traditionally discharged through smokestacks to the surrounding environment. Air-quality legislation in the United States, Sweden and Japan requires that the emission of sulfur oxides be controlled. A fluidized-bed combustor offers an elegant approach to solving this problem. Crushed limestone or dolomite is fed into the bed along with the high-sulfur coal. The sulfur in the coal combines chemically with the calcium in the crushed stone, to form solid calcium sulfate. Under suitable conditions and with suitable quality limestone, it is in this way possible to trap more than 90 percent of the sulfur, which remains with the solid ash and is discharged with it from the combustor.

Fluidized-bed combustion is a generic description, from which spring many lines of development and application. Design engineers can vary a system's size, operating temperature, operating pressure, bed depth, velocity of air flow, type and size of feed fuel, bed material, bed additives, heat-removal system, proportion of air to fuel, and other design characteristics. The choice of such design details depends on the application, as will be described in following sections.

Historical Development

The concept of the fluidized bed was first invented in the 1920s in Germany. The turbulent mixing and close contact of materials within a fluidized bed were found to promote chemical reactions. By the early 1940s, fluidized beds for cracking petroleum were a commercial success and other applications followed. As of the late 1950s, the fluidized bed was also well established as a technique for metallurgical heat treatment: a fluidized bed of fine sand heated by burning of natural gas was used to heat machined metal parts in a controlled fashion. Since then, American and other companies have sold many hundreds of such units, as well as fluidized-bed ore roasters, incinerators and similar units.

In such applications, however, the heat produced within the bed is eventually discharged as a by-product. The idea of using a fluidized bed actually to supply heat or to generate steam is of comparatively recent origin. From 1944 onwards, several companies in the United States and Germany developed designs for fluidized-bed systems to burn fuel in the bed to boil water and produce steam, but nothing came of these schemes. In France, a fluidized-bed boiler using two-stage combustion of coal (the "Ignifluid" boiler) was developed and was successful commercially, but attracted little attention. During the 1950s, emphasis in energy research and development shifted away from solid fossil fuel toward oil, natural gas and nuclear electricity. Technology for the combustion of coal, no matter how elegant or ingenious, elicited very little high-level interest and minimal financial support. This lack of interest made matters difficult for the handful of engineers pursuing their belief in the advantages of fluidized-bed energy systems.

Among this handful it is probably fair to give special mention to the late British engineer Douglas Elliott. At the end of the 1950s, when Elliott was working for Britain's Central Electricity Generating Board (CEGB), he in effect re-invented the coal-fired fluidized-bed boiler. The CEGB was primarily interested in large utility-size boilers for electricity generation, although at the time it too was losing interest in coal, in favor of oil and nuclear power. Nevertheless, Elliott's work at the CEGB's Marchwood Laboratories prompted further studies at the British National Coal Board's Coal Research Establishment at Stoke Orchard and at the Leatherhead Laboratories then operated by the British Coal Utilization Research Association (BCURA)(see National Coal Board profile).

Up to this point, it had always been assumed that a fluidized-bed boiler would operate at about atmospheric pressure. The only excess pressure in the system would be the small additional pressure needed to force the air up through the bed. In 1968, however, Elliott and his colleague Raymond Hoy of BCURA began to investigate the possibility of enclosing the fluidized-bed system inside a shell pressurized to several atmospheres, especially with a view toward using the hot pressurized combustion gases to turn the blades of a coal-fired gas turbine, and so turn an electricity generator. By 1969 they had built at BCURA an 8 MW (thermal) facility, which is still the world's largest operating pressurized fluidized-bed combustion test rig. Since that time, fluidized-bed systems have always been divided into the two sub-classes: "atmospheric" and "pressurized."

In the meantime, in the United States, in 1962, Michael Pope of the engineering-design firm Pope, Evans and Robbins (PER) (see profile) saw an operating fluidized-bed incinerator and was prompted to investigate the prospects for fluidized-bed energy systems. An exchange of visits between PER and the British fluidized-bed laboratories provided further encouragement. Beginning in 1965, with the support of the U.S. Office of Coal Research (OCR), PER built three atmospheric fluidized-bed test rigs. The largest of these, a 0.5 MW (thermal) unit at Alexandria, Virginia, started up in 1965. By 1970, the Foster Wheeler Corporation and the Combustion Power Company were carrying out experimental work on laboratory-scale fluidized-bed test rigs. However, the level of effort reflected little more than a holding action. Because oil and natural gas were still readily available and relatively inexpensive, market prospects for fluidized-bed combustion did not look promising.

Nevertheless, a wide variety of technical and engineering detail was studied: the phenomena of bed behavior, performance of structural materials, metallurgy, corrosion, heat transfer, combustion efficiency and so on. The high rate of heat transfer from the bed to the heat removal system suggested the possibility of "quenching": heat removal so rapid that it caused the bed temperature to fall, causing a loss of combustion efficiency and other unsatisfactory performance. Control of a fluidized-bed boiler also posed distinctive problems: "start-up," raising the bed temperature from room temperature until it would sustain combustion of fuel; and "turn-down," lowering the rate of heat production in the bed without allowing its temperature to drop too low to sustain combustion of fuel. Fuel-feed design also needed attention. It was necessary to establish the acceptable range of fuel-particle sizes, the best way to introduce them into the bed, and the number of feeding points required for a given area of bed.

Combustion Power-designed system pictured above burns wood waste to generate steam. The unit is a lineal descendant of experimental work begun in 1970. Photo: Courtesy Combustion Power Company, Inc.

The most intriguing group of problems, however, became of interest as a result of increasing concern about air quality. The potential of a fluidized bed for promoting chemical reactions showed promise as a means of reducing the emission of noxious gases from the combustion of fossil fuels. In 1968, under OCR sponsorship, PER's team at Alexandria, Virginia, fed limestone along with high-sulfur coal into their test fluidized-bed combustor and observed a lower sulfur dioxide content in the off gas. The creation of the U.S. Environmental Protection Agency (EPA) in 1970 led almost immediately to a joint research program between the EPA and the British National Coal Board, to study fluidized-bed combustion as a way of reducing the production of sulfur and nitrogen oxides from boilers. The program, completed in 1971, indicated that up to 95 percent of the sulfur in high-sulfur coal could be captured in a fluidized bed. However, British coal is in general low in sulfur, and this research attracted scant attention from British boiler users.

Both American and British research teams quickly established that calcium-containing minerals, in particular limestone and dolomite, reacted chemically in a hot fluidized bed with sulfur from coal, trapping the sulfur as solid calcium sulfate. But no one was quite sure why it worked. The effectiveness of the so-called "sorbent" stone varied with the temperature of the bed, the size of the sorbent particles, and the system pressure. Some research suggested that limestone was more effective in an atmospheric fluidized bed, while dolomite was more effective in a pressurized fluidized bed. Research was also initiated to ascertain the optimum proportion of the sulfur; beyond a certain level additional sulfur-trapping required disproportionally more sorbent, with diminishing effectiveness, while creating ever more solid waste. A lower so-called "calcium/sulfur ratio" or "Ca/S ratio" thus might not trap enough sulfur, but a higher ratio might lead to unnecessary use of sorbent and aggravation of the solid-waste problem. The optimum calcium/sulfur ratio appeared to be between 2.5 and 4 to 1, but varied considerably, not least between limestone and dolomite. As to the disposal problem, one view held that the ash combined with the additional sulfated sorbent would be significantly more difficult to dispose of than the ash alone. Another view held that the calcium sulfate was in fact useful, and could be spread on farm fields as a soil conditioner. Yet another view held that the sorbent stone should be treated to remove the sulfur, allowing the stone to be reused as a sorbent. This would reduce the amount of quarrying required to supply stone, but it would still leave the problem of disposing of the recovered sulfur in some form. Experiments at Argonne National Laboratory also indicated that sorbent stone lost its "sorbency" with repeated use.

Other problems, including control of particulates and of emissions of trace elements like sodium and potassium from fuel, were also identified; some remain unresolved. (Their environmental implications will be discussed in the next section.)

By the 1970s, a number of researchers felt ready to proceed to the pilot plant stage. In 1971, the British National Coal Board (NCB) proposed building a 20 MWe fluidized-bed power station near the NCB's Grimethorpe coal mine, but the government and the CEGB turned it down. In the United States, however, in 1972, the U.S. Office of Coal Research (OCR) agreed to provide 100 percent financing for Pope, Evans and Robbins to build a 30 MWe prototype atmospheric fluidized-bed power plant at Rivesville, West Virginia (see PER profile). In 1973, the OCR also funded the Combustion Power Company (see profile) to convert a 1 MWe pressurized fluidized-bed incinerator unit, the CPU-400, to coal-firing. Like Elliott and Hoy in Britain, the OCR researchers were interested in feeding the hot pressurized gases from a coal-fired fluidized-bed combustor into a gas turbine. The hot gases expanding through the turbine would turn a shaft which also carried an electric generator as well as a compressor to pressurize the combustor. The idea of a coal-fired gas turbine had indeed been tried in several countries not long after World War II. But the hot gases from the conventional firebox which was used carried too many particulates from the coal out of the firebox and into the turbine, wearing away -- "eroding" -- the turbine blades. Hoy and Elliott and the OCR researchers all considered the pressurized fluidized-bed combustor a much more promising basis for a coal-fired gas turbine. As a result of the fluidized-bed unit's lower operating temperature, fluidized-bed particulates were soft and relatively non-erosive, as opposed to the hard particulates from a conventional coal combustor. If a coal-fired gas turbine could be built, the hot combustion gases from the gas-turbine exhaust would still be hot enough to pass through a steam boiler; it would thus be possible to power both a gas turbine and a steam turbine from the same quantity of burning coal. Such "combined gas and steam turbine cycles" -- "combined cycles" for short -- were already available for the more expensive fuel oil usually burned in a gas turbine. But combined cycles would amplify considerably the usefulness of less expensive coal, making it possible to convert much more of the coal into electricity -- perhaps up to 50 percent in a combined-cycle plant, compared to 38 percent in an ordinary coal-fired plant equipped with a scrubber.

However, for a coal-fired closed-cycle gas turbine to be feasible, the problem of turbine-blade erosion and corrosion had to be overcome. The problem is caused by particulates. The classical way to remove particulates from combustion gases, used in conventional coal-fired plants, is the so-called "cyclone" collector which can remove upwards of 95 percent of the soot and dust in the gas stream. A high proportion of the remaining particulates can be removed by a more elaborate pollution-control device called an "electrostatic precipitator," which is likewise standard on all modern fossil-fueled plants. But control of turbine-blade erosion requires combustion gases to be almost entirely free of particulates; even removal of more than 99 percent, using cyclones and precipitators, may not suffice. Work on hot-gas clean-up is still a major area of research for fluidized-bed systems. In addition, some researchers express concern about turbine-blade corrosion resulting from the sodium and potassium in coal.

Progress on atmospheric systems has been more rapid. In Britain, the National Coal Board as a partner in the newly-established Combustion Systems Limited (CSL) (see profile) agreed in 1974 to convert a boiler at the Babcock & Wilcox, Limited (B & W Ltd.) factory in Renfrew, Scotland to fluidized-bed firing (see B & W Ltd. profile). The 40,000 pound-per-hour Renfrew boiler started up in August 1975, and for two years was the largest operating fluidized-bed boiler in the world. It operated so successfully that in August 1976 B & W Ltd. announced that they would offer units capable of producing up to 500,000 pounds of steam per hour commercially, with a warranty.

Meanwhile, elsewhere in Britain, a radically new approach to fluidized-bed systems was taking shape. Douglas Elliott, who now held the post of professor in the Department of Engineering at Aston University in Birmingham, had taken out patents on systems using fluidized beds much shallower than usual (a few inches as compared to a foot or more). In 1972, Elliott and his partner Michael Virr founded a new company called Fluidfire Development, Ltd. (see profile) to develop and market this and other concepts. Fluidfire sister companies rapidly gained success marketing gas-fired fluidized-bed heat-treatment furnaces and incinerator-boilers, and a novel design of fluidized-bed heat exchanger to extract useful heat from flue gases. In 1977, a commercial model of the fluidized-bed heat exchanger was installed on a 40,000-ton Norwegian tanker. Fluidfire also began development of small fluidized-bed systems for domestic use.

Perhaps the single most important energy-related event of the 1970s, however, was the OPEC oil embargo and fourfold price rise of 1973, which radically changed both government priorities and energy economics. Though not immediate, the impact on fluidized-bed technology was eventually very important. In the United States in 1974, the Office of Coal Research (OCR) was absorbed into the new Energy Research and Development Administration (ERDA). ERDA took over the existing OCR research and development programs including the Pope, Evans and Robbins' Rivesville project (see profile). Coal research at ERDA originally emphasized coal gasification and liquefaction (see ENERGY FUTURES: Industry and the New Technologies). The 1975 ERDA budget contained $109.8 million for gasification and $94.7 million for liquefaction, but only $19.8 million for fluidized-bed combustion. However, by late 1975, interest in the direct combustion of coal was increasing. In September of that year, ERDA issued "Program Opportunity Notices," invitations to organizations to join in research and development of direct coal combustion technologies, particularly fluidized-bed combustion. Reflecting the fact that, of all fluidized-bed applications, atmospheric units designed for industrial use had the best near-term commercial prospects, ERDA awarded contracts to Battelle Memorial Institute, Combustion Engineering, Inc., Exxon Corporation, Fluidyne Engineering Corporation, and Georgetown University for demonstration industrial atmospheric fluidized-bed units.

ERDA was also, however, interested in the technically more complex task of developing large pressurized fluidized-bed/gas-turbine units for electric utilities to use in power generation. In order to develop an adequate data base for further work, ERDA signed a contract in March, 1976 with the Curtiss-Wright Corporation to design and construct a pilot coal-fired combined-cycle plant using a pressurized fluidized-bed combustor. Contracts were also signed with Burns and Roe, Inc. and the General Electric Company for pressurized fluidized-bed designs and technology development.

These atmospheric and pressurized fluidized-bed contracts joined an ERDA program which already included the Pope, Evans and Robbins' Rivesville project, an Oak Ridge National Laboratory investigation of a coal-fired "total energy system" utilizing an atmospheric fluidized-bed combustor, Combustion Power Company's CPU-400 pressurized fluidized-bed/gas-turbine test facility, and various support studies at Argonne National Laboratory in Illinois and the Morgantown Energy Research Center in West Virginia.

On March 1, 1977, President Carter proposed a Department of Energy to provide a framework for carrying out a national energy policy. The Department of Energy Organization Act was signed into law on August 4, 1977, and the next day veteran cabinet member James Schlesinger became the new department's first Secretary.

The Department of Energy (DOE) subsumed ERDA and its fluidized-bed combustion program. Under DOE, several projects conceived by the OCR and ERDA are coming to fruition. Besides the industrial demonstration projects and the utility pilot projects mentioned above, atmospheric and pressurized "Component Test and Integration Units" (CTIUs) are being constructed at the Morgantown Energy Research Center and the Argonne National Laboratory, respectively. These units, "a research and development guy's dream" in the words of Dr. Steven Freedman, director of DOE's fluidized-bed program, are designed to provide flexible test units for studying every aspect of and problem associated with fluidized-bed combustion. DOE is also funding a new atmospheric fluidized-bed program for facilities to burn anthracite waste and is considering another to burn lignites.

Some fluidized-bed experts surveyed by INFORM criticize DOE for putting too much of its $56 million 1978 budget for fluidized-bed research into the CTIUs and the many support studies which it funds. They feel the commercial prospects of fluidized-bed units, particularly those being developed for utility use, would be heightened if DOE would fund one or more large demonstration fluidized-bed plants. Without these demonstration plants, the argument goes, utilities will not develop the adequate confidence in fluidized-bed technology they require to invest their own money in such a system. DOE, while investigating the possibility of funding a demonstration plant, feels it important to fully understand and resolve the problems associated with fluidized-bed technology before proceeding to the demonstration plant size. DOE has been doing preliminary planning for a 200 MWe utility demonstration plant since 1976 and hopes to have sufficient funds to start construction of such a facility in 1980.

Internationally, the 1973 oil price rise also had far-reaching aftereffects. One was the establishment of the International Energy Agency (IEA) of the Organization for Economic Cooperation and Development (OECD). The OECD/IEA at once set up a joint program of coal technology cooperation, whose major project was construction of an 80 MW (thermal) pressurized fluidized-bed test installation at Grimethorpe, Yorkshire, to be managed by a team from Britain's National Coal Board (see profile on NCB [IEA Services]). The Grimethorpe project was sponsored in partnership by the British, United States, and West German governments; the unit is to be commissioned in 1979.

In Sweden, the ability of fluidized-bed combustion systems to burn virtually any fuel led the town of Enkoeping, near Stockholm, to order a 25 MW (thermal) fluidized-bed boiler for the town's district heating system (see profile on AB Enkoepings Vaermewerk). The Enkoeping boiler came on stream in February 1978, and has been in successful operation since that time, burning heavy oil and subsequently coal. The Swedish turbine manufacturer Stal-Laval Turbin AB has also been taking an interest in coal-fired fluidized-bed systems for industrial co-generation and district heating. In 1976, the British subsidiary of Stal-Laval (see profile) teamed up with Babcock & Wilcox, Limited (U.K.) (see profile) to design a 170 MWe coal-fired fluidized-bed combined-cycle power plant for the American Electric Power Company (see profile). Babcock & Wilcox, Limited also proposed a similar unit for British Columbia Hydro in Canada, and both proposals are progressing in 1978 toward definite decisions on orders. In 1977, Babcock & Wilcox, Limited also won three contracts from the state of Ohio for coal-fired atmospheric fluidized-bed boilers, two of industrial size and one of utility size; a fourth Ohio order is planned (see Babcock & Wilcox, Limited profile).

The 30 MWe Rivesville plant (see Pope, Evans and Robbins profile) started up in August 1977, becoming the world's largest operating fluidized-bed boiler, although it encountered some difficulties, especially with coal feeding. In Britain the National Coal Board (NCB) at last sidestepped the CEGB's lack of interest in fluidized-bed systems, and contracted to convert an 80,000 pound-per-hour boiler at the River Don factory of the British Steel Corporation to fluidized-bed coal-firing. The River Don unit is to be commissioned in 1978. The NCB also contracted for a series of other smaller industrial fluidized-bed conversions, some of which are already in operation (see NCB profile).

Meanwhile, the research and development work continued. Battelle Columbus Laboratories in the United States and Fluidfire in Britain came up with second-generation designs of atmospheric fluidized-bed combustors capable of much larger heat outputs per unit size (see profiles on Battelle and Fluidfire). Interest in fluidized-bed energy technology began to burgeon elsewhere. In West Germany, the mining research organization Bergbau-Forschung undertook design of a utility-size pressurized fluidized-bed unit. In India, the major engineering firm of Bharat Heavy Electricals, Ltd. developed their own design of fluidized-bed system. Work on fluidized-bed combustion also began to attract support in the Netherlands, Denmark, Ireland, Japan and other countries. There seemed to be international recognition that fluidized-bed energy technology, after its long period of neglect, was at last coming in from the cold.

Environmental Implications

The extraction and utilization of coal has a dishearteningly wide variety of environmentally unpleasant consequences. Deep mining may cause sinking of the land surface above, and expose miners to extreme health and safety risks. Surface mining may leave long-lasting scars on the earth, although European mining practice has long been more scrupulous about restoration of surface-mined areas than has the general United States practice. The burning of coal introduces a number of products into the atmosphere, including carbon dioxide, nitrogen and sulfur oxides, and soot and ash. Any one of these warrants concern. The buildup of carbon dioxide in the earth's atmosphere, for example, has recently begun to attract serious attention because it may interfere with the mechanisms which govern the global climate. It is not possible to say precisely how much of the carbon dioxide buildup has resulted from the burning of coal; the destruction of forests, which absorb this gas, has also had an impact. But the burning of coal invariably releases additional carbon dioxide gas in the air.

At the very least, it is therefore environmentally advisable to obtain as much useful energy as possible from each ton of coal mined and burned. Fluidized-bed systems appear well suited for applications which make more efficient use of coal, such as combined cycle systems and co-generation of electricity and heat. Fluidized-bed systems are also able to burn a much wider range of coals than conventional systems, as well as other, less combustible materials. It is thus possible to be more versatile in exploitation of fuel resources, using fuels that are locally available and thus more economical.

Certain problems associated with the use of coal will remain, even in the context of fluidized-bed combustion systems. Coal delivery and stocking will raise difficulties. But coal to be burned in a fluidized bed does not require washing or pulverization; this reduces some of the problems of water pollution and occupational hazards which arise in traditional coal-burning systems.

However, the most distinctive environmental advantage associated with fluidized-bed systems is undoubtedly the possibility of comparatively direct and simple control of noxious combustion gases. At the combustion temperature in a conventional fossil-fuel boiler, a significant proportion of the nitrogen in the combustion air combines with oxygen to form nitrogen oxides (NOx). Though certain kinds of boiler modifications can cut down on this reaction, the remaining nitrogen oxides are discharged to the atmosphere in the stack gases. Nitrogen oxides are themselves variously toxic; they are also a major constituent of smog. In the United States, present air-quality laws limit the maximum allowable emission of nitrogen dioxide (NO_2) from a new coal-burning plant to 0.7 pounds per million Btu of heat and the U.S. Environmental Protection Agency is considering making this standard stricter.

The lower operating temperatures possible in a fluidized-bed boiler, however, are accompanied by a substantial reduction in nitrogen-oxide production. Research at the Argonne National Laboratory in the United States and at the National Coal Board in Britain indicates that atmospheric and pressurized fluidized-bed systems will both be able to comply with United States requirements for nitrogen-oxide emissions. There may be some operating constraints on atmospheric systems to achieve compliance; pressurized systems appear to produce significantly less nitrogen oxide than atmospheric, other conditions being equal. The pressure-dependence effect is not as yet fully understood.

Sulfur emissions are another potential environmental problem when sulfur is present in boiler fuel. The burning of coal containing sulfur will result in the formation of sulfur oxides. Sulfur dioxide and trioxide are known to aggravate respiratory ailments, to turn surface waters acid, and to create other environmental problems. In consequence, United States laws now require that new coal-burning utility plants emit no more than 1.2 pounds of sulfur dioxide (SO_2) per million Btu of heat. As of spring, 1978, the U.S. Environmental Protection Agency (EPA) was considering a proposal to require control of 85 percent of the sulfur in any coal being burned, with an upper limit of 1.2 pounds per million Btu and a lower limit of 0.2 pounds per million Btu. These limits mean that if capturing 85 percent of the sulfur in the coal causes emissions to drop to less than 0.2 pounds of sulfur per million Btu, the plant is required to remove only such a percent as to get emissions to 0.2. If removing 85 percent leaves emissions of over 1.2 pounds per million Btu, the plant is required to increase its percentage removal until the 1.2-pounds-per-million-Btu limit is reached.

Compliance with EPA requirements is currently presenting most United States coal users with knotty technical and economic difficulties. Sulfur can be removed from coal before burning, but this process is expensive and thus far cannot remove sufficient amounts of sulfur to meet the considered regulations by itself. The technical option now most prominently pursued is removal of sulfur oxides from flue gases after burning. But flue-gas desulfurization, by means of so-called "scrubbers," leaves much to be desired. Scrubbers themselves are massive and expensive capital installations, accounting for a significant fraction (15 percent) of the total investment for a new coal-burning plant. Some companies claim that scrubbers are proving unreliable and awkward to maintain. Their greatest drawback, however, may be that they produce large quantities of wet sludge as a waste product. Disposal requires substantial areas of land which then become sterile and useless for the foreseeable future. It is clear that a less troublesome alternative way to control sulfur oxides from coal would be welcome. For this purpose especially, fluidized-bed systems look more and more promising.

As mentioned earlier, the fluidized bed can trap sulfur from a fuel directly during the process of burning the fuel. While the carbon in the fuel is reacting with the oxygen from the air -- that is, burning -- the sulfur in the fuel combines not only with the oxygen but also with the surface layers of any solid particles in the bed. Research has shown that certain minerals, particularly limestone and dolomite, are very much more effective sulfur-traps than other substances. Limestone usually consists almost entirely of calcium carbonate. Dolomite consists of roughly a half-and-half mixture of calcium and magnesium carbonates. In either case, the particles of the mineral in the bed, when raised to the overall temperature of the bed in the presence of sulfur, undergo a series of chemical changes: some of the carbonates are broken up and replaced by sulfates. In this manner, it is possible to capture over 90 percent of the sulfur in the coal. However, research has established that optimum trapping depends on the proportion of calcium to sulfur. The usual procedure is to premix the coal with crushed stone in a suitable proportion, and feed the mixture into the fluidized bed. As noted earlier, the optimum proportion of calcium to sulfur appears to be between 4 and 2.5 to 1; research is still continuing. The optimum also varies with operating temperature and pressure, the quality of the "sorbent," and the size of sulfur-trapping particles. Clearly, the less additional stone that must be added the better. Quarrying the stone, crushing it, and delivering it to the plant all involve potential environmental problems which must be set against the sulfur-removal advantage. They may also occasion considerable costs. Utility representatives surveyed by INFORM stressed that from an economic point of view, a calcium-to-sulfur ratio of about 1.5 to 1 would be very desirable.

There then remains the question of what to do with the waste created by this sulfur-removal process, a solid granular material which includes both coal ash and sulfated sorbent. Ash disposal has always been a problem for coal users. For high-sulfur coals, this sulfur-removal technique may more than double the volume of solid waste. This increase must obviously be debited against the sulfur removal. It is nevertheless generally agreed that compared to the viscous, liquid sludge from scrubbers, the sulfated stone from a fluidized bed -- a dry, innocuous powder mixed with ash -- is harmless and manageable. Some developers, Pope, Evans and Robbins, for example, even hold the view that the calcium and magnesium sulfate are useful soil-conditioning agents, and can with advantage be spread on farmland in the vicinity of the plant. Farmers have used "land plaster" for years to replace the sulfur taken up from the soil by crops. Tests show that spent bed material is as effective in this respect as commercial land plaster and no detrimental effects have been found.

In the future, it may be feasible to reduce this solid waste even further. It may be possible to use the stone merely as a carrier for the sulfur, and to remove the sulfur from the stone after it has been discharged from the fluidized-bed system. Reuse of stone would of course also minimize quarrying, crushing, and transport. Laboratory work on this so-called "sorbent regeneration" has been underway for some time, but many problems remain unresolved. Regeneration of the stone thus far appears to be a chemical process at least as complex as the original sulfur-trapping, with sulfur-recovery requiring considerable additional facilities. Furthermore, at this stage it is not clear whether the sulfur produced will be economically useful in the quantities anticipated. The relative merits of once-through use of stone, followed by direct disposal, versus reuse of stone, followed by recovery of sulfur, are as yet difficult to judge. Work on both options is continuing.

It must be added that the relevance of sulfur-trapping as a virtue of fluidized-bed systems has long been disputed by Britain's Central Electricity Generating Board (CEGB). The present CEGB view is that for its purposes better sulfur-control is unnecessary, since tall stacks provide adequate dispersal of sulfur oxides created by burning British coal, which is in general low in sulfur. However, the dispersed sulfur oxides generated in Britain do eventually return to the ground, and the ground they return to may be in Scandinavia. The Scandinavian governments have expressed sharp dissatisfaction to the CEGB at the acidification of their waterways. The issue is far from settled.

The CEGB also considers that a fluidized-bed system would still require a tall stack to provide dispersal of the residual sulfur dioxide not removed by the bed. There is, however, no published legal requirement about sulfur-oxide emissions in the United Kingdom. The former Alkali Inspectorate, now subsumed into the Health and Safety Executive of the British Government, carries out all its regulatory functions in private discussion with the industries it is regulating. Industry is called upon to use "best practicable means" to minimize emissions. But the judgment as to practicability and as to compliance remains a matter for the Inspectorate and the industry concerned. There is, it should be said, much dissatisfaction in the United Kingdom about this arrangement, but efforts to change it have thus far been unavailing.

The burning of coal also produces particulates, that is, dust and smoke. In a fluidized-bed system these particulates are formed in the bed. The rate at which they are carried upwards out of the bed depends on the velocity of the fluidizing air. At the velocities usually employed, the surface of the bed appears to behave rather like the surface of a liquid, impeding most particulates from rising above the bed. However, some particulates are carried into the stack, where traditional pollution-control devices -- cyclones and baghouses -- can remove almost all of them. Fine particulates which escape these devices and are emitted into the atmosphere, are beginning to worry many environmentalists. Their effects are unknown and research into this area has only recently gotten underway.

Another problem presented by particulates is not an air-quality problem, but rather that of turbine-blade erosion, if the combustion gases are fed into a gas turbine. As mentioned earlier, cyclones and precipitators may not clean up combustion gases adequately for this purpose. Nevertheless, those working on coal-fired gas-turbine applications point out that if they can clean up gases enough to suit turbine blades, the remaining particulates in gases will certainly present no significant environmental problem.

Research is also being undertaken on other air pollutants released during the burning of coal, in particular trace elements and organic compounds. Preliminary results indicate that such emissions may be more readily controlled in fluidized-bed systems than in conventional fossil-fired systems. But further work will be needed.

The U.S. Environmental Protection Agency (EPA) is conducting a $4 million program to identify potential environmental problems associated with fluidized-bed combustion, in order to assess the available control technology and to integrate the technology into the fluidized-bed research and development process. The organizations doing this work for EPA include the Battelle Memorial Institute, Argonne National Laboratory, and Westinghouse Research Laboratories. An EPA quarterly newsletter, FBC Environmental Review, is being used to report on the program's progress to interested parties.

In any event, if fossil fuel is to be burned at all, research seems to indicate that the emissions can be more successfully minimized in a fluidized-bed system than otherwise.

Prospects For Commercialization

Energy-technology research and development aims to devise ways to supply and use energy as cleanly, as cheaply and as controllably as possible. Of all the available fossil fuels, coal is the most plentiful. It is not, however, easy to use cleanly, cheaply or controllably. The rebirth of interest in coal has been accompanied by new efforts to overcome these difficulties.

One of the most promising lines of technical development is that based on the fluidized bed. The possibilities of this development have been recognized for at least two decades. But early work on fluidized-bed systems was done in a climate of dwindling interest in coal. Oil and natural gas were more manageable, and appeared abundant; nuclear energy received virtually all the remaining research and development support. If coal was to decline in importance, there was little reason to look for ways to overcome its disadvantages.

The subsequent change in the climate of opinion has been dramatic. Energy planners all over the world, whatever their other differences, now look to coal as the one essential fuel, at least for the foreseeable future. The technical and economic context has been transformed. The potential for fluidized-bed systems, which at one time could be dismissed by all but a handful of stubborn believers, is now generally recognized by those familiar with the technology -- as yet, to be sure, a somewhat limited group. How, and at what rate, the potential for fluidized-bed energy technology can be realized is still uncertain. Before surveying the areas of uncertainty, however, this commentary must stress one point. The problems of fluidized-bed systems must be differentiated from those which are endemic to coal itself. All but the most single-minded coal enthusiasts will concede that, given the choice, oil and natural gas are more satisfactory fuels than coal, and that solar energy for appropriate applications would be even better. The interest in advanced coal technologies starts from the premise that the choice of oil or natural gas is no longer reliably available, that nuclear power carries high environmental risks and provokes strong public opposition, and that solar technology is as yet not adequately available or economic for certain important applications, particularly those requiring intense heat, such as certain high-temperature industrial operations. The increasing use of coal, especially in the medium term, seems likely to be inevitable. If this is so, the potential advantages of fluidized-bed systems warrant serious study by all those involved in energy use and supply decision-making.

Present uncertainties about fluidized-bed systems fall into several categories: technical, economic, environmental and political. Opinions among fluidized-bed specialists differ widely, and indeed sometimes acrimoniously, on a number of key aspects. But all agree to begin with, that the basic concept of fluidized-bed combustion is sound and that fluidized-bed combustion works. As noted earlier, three American, one Norwegian, and three British companies are offering fluidized-bed combustors for small-scale industrial use on a commercial basis, and a number of such systems are functioning. When it comes to development of larger, more efficient or more sophisticated systems, however, opinions tend to diverge, with each group convinced that its work is on the right track, and each group more or less doubtful about the work of others in the field.

The most fundamental divergence is between those who feel that others are trying to proceed too fast, and those who feel that others are dragging their feet. In general, opinion among the United States fluidized-bed community is that Europeans are trying to proceed too fast, while European opinion is that the United States is dragging its feet. There are of course degrees and exceptions to this general observation, but it is one which American decision-makers in particular might wish to investigate further. They may be receiving from their fluidized-bed advisors an unduly tentative or even pessimistic view which is not shared outside the United States.

Another fundamental divergence of opinion, more technical in nature and not so absolute, is between atmospheric and pressurized systems. Some companies, like PER (see profile), are concentrating entirely on atmospheric systems; others, like Curtiss-Wright (see profile), entirely on pressurized systems. Still others, like Babcock & Wilcox, Limited (of Britain) (see profile), are actively involved in both lines of development. Companies pursuing both lines tend to feel that atmospheric fluidized-bed combustion will be preferred for industrial-size boilers, and that utilities will prefer pressurized systems, which can be adapted to combined-cycle use. For a larger utility-size plant, the capital-cost savings realized by the pressurized fluidized-bed combustor's smaller size for the same amount of output will be important, as will the improved efficiency achieved through combined cycles.

The divergence of opinion as to how fast to proceed, and what lines of development to pursue, may also be related to differing perceptions of the difficulty of scale-up, and of the usefulness of data from pilot-scale test rigs in designing full-scale commercial units. Research laboratories in the United States, for instance those now constructing the Component Test and Integration Units, perhaps understandably consider it advisable to carry out exhaustive investigation, on a small scale, into emissions and their control, sorbent behavior, and other fluidized-bed phenomena. However, some companies, especially those in Europe, tend to insist that small-scale results will not necessarily prove applicable to scaled-up commercial units. Such companies, including Stal-Laval (of Sweden) and Babcock & Wilcox, Limited (of Britain) consider that enough data already exists -- not the least from the Babcock & Wilcox, Limited, Renfrew unit -- to permit the construction of major prototype facilities, and that the factors involved in a scale-up from small size will make the precision of future small-scale results illusory. An intermediate opinion is held by those involved in the Grimethorpe, England, test rig, which will be of prototype size but used purely for research (see NCB [IEA Services] profile). The Grimethorpe rig will, they say, be more flexible than a working commercial unit and lend itself better to necessary full-scale experimentation. They doubt that there is enough relevant data available for Curtiss-Wright and American Electric Power/Stal-Laval/Babcock & Wilcox, Limited (see profiles) successfully to design and operate full-scale pressurized fluidized-bed combustor/gas turbine systems. The Grimethorpe team considers it premature to tackle both the problem of pressurized fluidized-bed combustion itself and the problem of coupling such a combustor to a gas turbine. On the other hand, the Stal-Laval/Babcock & Wilcox team feels that the Grimethorpe rig will be "reinventing the wheel," carrying out research on a non-commercial unit which might perfectly well be done by building a commercial prototype and acquiring the desired experience with it in actual service. Some American experts feel that the Stal-Laval/Babcock & Wilcox team are overlooking the problems of gas turbine-blade corrosion due to sodium and potassium in the coal, an area which they feel is worthy of consideration.

In turn, all of the groups involved in pressurized systems for utility use consider the Rivesville plant -- an atmospheric system designed for utilities -- effectively already obsolete, and even a possible embarrassment. They point out that it was conceived and designed (by Pope, Evans and Robbins and the Office of Coal Research) under very different economic criteria six years ago, before the 1973 oil price rise. The objective at that time was to produce a system that would be competitive with then-cheap oil. The critics feel that the problems the Rivesville plant is now encountering are in part a consequence of the tight specifications laid down when the objective was to build a stringently cheap boiler. Today, the Department of Energy (whose fluidized-bed combustion program is a lineal descendant of the Office of Coal Research's) is primarily concerned with sulfur-oxide emissions control. Fluidized-bed combustors must compete economically not with oil-fired boilers, but with conventional coal-fired boilers using scrubbers. Pope, Evans and Robbins (PER) concedes that times and priorities have changed. However, PER questions a wholesale switch to pressurized fluidized-bed combustion. PER believes that the costs saved by smaller size for the same output will be offset by the additional cost of coping with the higher pressure, especially as regards feeding coal and removing solid waste. Michael Pope of PER considers that atmospheric systems will be preferable at all sizes, from small industrial right up to full-scale utility boilers.

Henrik Harboe of Stal-Laval believes the opposite: that pressurized systems may be preferable for medium-to-large-scale applications. He cites the advantage of being able virtually to double a pressurized system's useful energy output by using combined cycles or co-generation, and says this will be the most effective lever to persuade industries to convert their boilers to pressurized fluidized-bed coal firing. Art Fraas, a consultant to the Oak Ridge National Laboratory (ORNL), points out that using ORNL's closed-circuit hot-air turbine system (see profile) makes it possible to combine both atmospheric fluidized-bed combustion and combined cycles or co-generation.

Dr. Steven Freedman, director of DOE's fluidized-bed program, sees industrial-size atmospheric fluidized beds as having the nearest-term commercial availability. These systems' greatest advantage, he feels, is their fuel-use flexibility. Freedman sees utility systems as a longer-term proposition. DOE hopes to have both a large atmospheric and a large pressurized fluidized-bed plant on-line by 1984. Freedman says the object is to allow utilities to choose between the two systems. He sees advantages to both approaches: atmospheric being simpler and more reliable while pressurized has a slightly higher efficiency and slightly lower sulfur-dioxide emissions.

There are more specialized differences of opinion. Battelle Memorial Institute (see profile) and other United States fluidized-bed researchers have expressed concern about possible corrosion of boiler tubes. Others, especially in Europe (including Combustion Systems Limited, Babcock & Wilcox, Limited, and Stone-Platt Fluidfire Limited [see profiles]), have seen nothing to suggest that such problems will arise, and consider them non-existent. Coal feed has given persistent trouble at the Rivesville unit, and several other United States teams are also concerned about feed problems. But European teams have seen no such problems. The general European view seems to be that coal can be fed from above the bed, within or under it, and that the easiest way, from above, is quite satisfactory. Coal fed from above remains within the bed long enough to burn completely, without undue carryover of unburned particles into the flue gases. The European view is that if, when moving to large-scale applications, a company uses a special carbon burn-up cell to burn unburned particles (as at Rivesville), this indicates that the design of the main bed is less than optimum.

Still another divergence of technical opinion is between those who favor cell and bed dimensions small enough to permit prefabrication in the manufacturer's shop, and those who see no special advantage in this, favoring instead beds tens-of-meters along a side for large-scale applications. The smaller-scale or modular approach offers the advantages of replication, and also of stable working conditions in an established shop, as against site

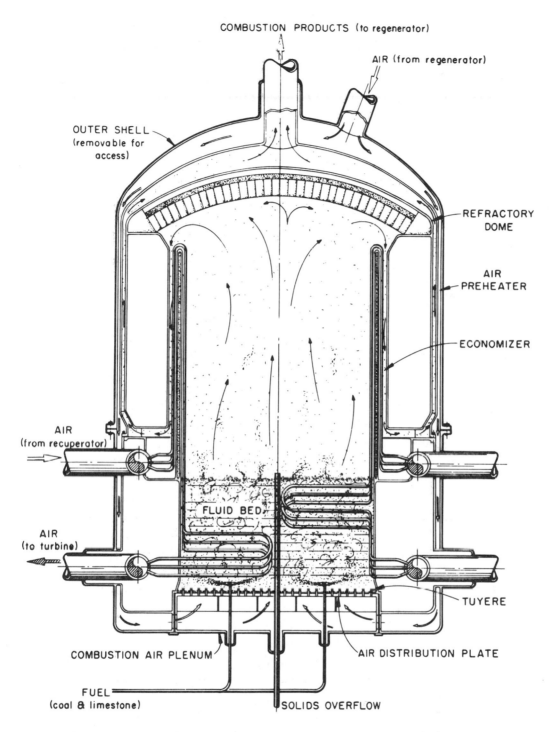

COMBUSTION PRODUCTS (to regenerator)

AIR (from regenerator)

OUTER SHELL (removable for access)

REFRACTORY DOME

AIR PREHEATER

ECONOMIZER

AIR (from recuperator)

AIR (to turbine)

FLUID BED

TUYERE

COMBUSTION AIR PLENUM

AIR DISTRIBUTION PLATE

FUEL (coal & limestone)

SOLIDS OVERFLOW

SCHEMATIC DIAGRAM SHOWING FLUIDIZED-BED COAL-COMBUSTION SYSTEM DESIGNED TO SERVE AS A HEATER FOR A CLOSED-CYCLE GAS TURBINE

This Oak Ridge National Laboratory design is unusual because it uses an atmospheric rather than a pressurized fluidized-bed combustor in combination with a gas turbine. In this system, the gas which drives the turbine is heated in tubes in the fluidized bed. Drawing: Courtesy Oak Ridge National Laboratory.

work. However, Babcock & Wilcox, Limited, for instance, considers that the modular design may not be as flexible in actual operation as the custom-built large-scale bed, which may be subdivided into separate cells with separate air and fuel feed and other controls.

Such technical and procedural disagreements should not be taken to indicate any deep antagonism between the different camps. On the contrary, the reason for the disagreement is virtually always a feeling that someone else is following a pathway which may lead to failure, and thereby cast a shadow over the whole of fluidized-bed technology. Even where disagreements exist, there appears generally to be a prevailing hope that both sides will be vindicated, and that the differing approaches will all succeed or prove valuable for different applications.

Uncertainties within the fluidized-bed community itself are compounded by uncertainties arising from external factors. There is, to begin with, the basic economic issue. How much energy supply will be required? How much of it will be provided by coal? How will the cost of coal compare over time with the cost of other fuels? What proportion of energy needs will be met by electric utilities? What proportion will be met directly by industry and commerce, or by the domestic sector? How much will new investment cost? How will the investment cost of a given fluidized-bed system compare with other fluidized-bed systems, or with other coal-fired systems? How much will a fluidized-bed retrofit cost, compared with retrofitting of scrubbers?

The role of government financial and fiscal policy is crucial, and at present unpredictable. It gives rise to another set of uncertainties. Will there be government grants or tax breaks given for installation of fluidized-bed systems? If there is a possibility of grants or tax breaks in the future, would it not be better to wait until then to invest rather than to lay out the money now? The United States government has approved low-interest bonds for pollution-control investment. Such bonds may be used to finance retrofitting of scrubbers, but not to finance construction of new fluidized-bed systems, which do not emit the pollution in the first place. At the very least, it would seem sensible to make such bonds available also for conversion of existing industrial units to coal-fired fluidized-bed firing.

A recent publication of the Congressional Research Service, "Energy Policy As If It Really Mattered," points out that the United States government could advance the commercialization of fluidized-bed combustion by installing systems in its numerous industrial and administrative facilities. The state of Ohio is doing just that; the Ohio Energy Resource and Development Agency (OERDA, now the Ohio Department of Energy, ODOE) has ordered an atmospheric fluidized-bed boiler for a state penitentiary and a fluidized-bed retrofit of a boiler at a state mental hospital from Babcock & Wilcox, Limited (U.K.). A third fluidized-bed boiler has been ordered which ODOE hopes to lease to the Columbus and Southern Ohio Electric Company. ODOE is also funding part of the development program at Battelle Memorial Institute (see Babcock & Wilcox, Limited [U.K.] and Battelle Memorial Institute profiles).

Ohio took the initiative in the fluidized-bed field because it has large reserves of high-sulfur coal. Illinois, with similar large high-sulfur-coal reserves, has also taken steps to encourage coal utilization. The Illinois Division of Energy, part of the state's Department of Business and Economic Development, has a $65 million bonding authority which is designated for "coal development within the state of Illinois." Of these Coal Development Bond Funds, $750,000 has been allocated to the 50,000 pound-per-hour industrial atmospheric fluidized-bed combustion unit being constructed by Combustion Engineering, Inc. at the Great Lakes Naval Training Station in Illinois (see Combustion Engineering, Inc. profile).

Other state organizations are providing support for fluidized-bed projects. The Iowa Energy Policy Council is involved in a fluidized-bed gasification plant to be constructed in Forest City, Iowa, which will use a fluidized-bed combustor in the gasification process, and an atmospheric fluidized-bed air heater being constructed by the Fluidyne Engineering Company for the Owatonna Tool Company, which is specifically designed to burn Iowa coal (see Fluidyne Engineering Corporation profile). The New York State Energy Research and Development Authority (NYSERDA) sponsored a Combustion Engineering study of the feasibility of an atmospheric fluidized-bed retrofit of Consolidated Edison's 500 MWe Arthur Kill #3 plant on Staten Island, a borough of New York City. NYSERDA is also helping to fund laboratory work at the General Electric Company on hot-gas clean-up and other problem areas associated with fluidized-bed combustion. NYSERDA sees fluidized-bed combustion as an effective way to bring the burning of coal back to the cities, and is particularly interested in district heating systems utilizing fluidized-bed combustors.

In the United States, the energy problem affords state and local governments an opportunity to gain a greater measure of control over the policies affecting the lives of their inhabitants. Each state is in the position to formulate an energy program utilizing technologies appropriate to its own combination of conditions and resources. Particularly for states with high-sulfur coal reserves, fluidized-bed combustion units could be an important piece in the appropriate technology mosaic.

One further difficulty cited by many companies interested in developing utility-size fluidized-bed combustion units is uncertainty over environmental standards. Regulatory and environmental standards, still relatively new phenomena, are in flux everywhere. Standards, when proposed, are promptly challenged, often through litigation, by industry, environmentalists, or both. Investors and plant manufacturers state that they are unable to proceed with confidence for fear that a system which meets present standards will ere long be confronted with tighter standards it can no longer meet. As noted earlier, the U.S. Environmental Protection Agency is considering proposing a standard in 1978 which requires removal of 85 percent of the sulfur in any type of coal, to an upper limit of 1.2 pounds per million Btu and a lower limit of 0.2 pounds per million Btu. Some United States fluidized-bed teams express uneasiness about the most restrictive standards; but Babcock & Wilcox Limited (U.K.) representatives point to their own sulfur-control results which they have no doubt demonstrate the feasibility of compliance with even this latest stringent standard. Ironically enough, the sulfur-control policy in Europe is even more uncertain than that in the United States; indeed, in Britain it is presently effectively nonexistent.

Other question marks, common to all technical innovations, also hang over fluidized-bed systems. These can be combined into the single question of warranties. If designers and builders are adequately confident that their fluidized-bed systems will perform as intended, they must be prepared to back their confidence with a warranty. As of April, 1978, only Babcock & Wilcox, Limited (U.K.), Energy Equipment Company Limited, Energy Resources Company, Foster Wheeler and Mustad and Søn offer warranties for fluidized-bed combustion units (see profiles and appendix). But the picture is changing rapidly. Few fluidized-bed combustion people doubt that the warranty problem will soon be resolved.

Prognosis

There is general agreement both in the United States and in Europe that atmospheric fluidized-bed boilers will be the first to establish themselves in the market, meeting the needs of industry for power and steam. Observers surveyed by INFORM feel this will almost certainly come about within the coming decade and possibly sooner, particularly as conversions and retrofits. Pressurized systems of all sizes are expected to be delayed somewhat longer; again, the smaller sizes, from industrial up to small utility and co-generation uses -- less than 200 MWe -- are likely to make a mark first, perhaps by 1990. Full-scale utility boilers will take the longest to arrive, for several reasons. Utilities are reluctant yet to undertake the investment and commitment necessary. They are presently faced with cash-flow problems and an existing superfluity of base-load generating capacity, both in the United States and Europe. Future growth in electricity use is uncertain. However, the Electric Power Research Institute, the research arm of the utility industry, is spending $6 million in 1978 to investigate barriers to utility use of fluidized-bed combustion. Two factors in which utilities especially want to see improvements are calcium/sulfur ratios and combustion efficiency. The scale-up of either atmospheric or pressurized fluidized-bed combustion to a size presently considered appropriate for utility use is still a significant leap beyond present experience, and hesitance is probably justified. But the idea that utility plants must be large central units is steadily changing, and the rapid accumulation of operating experience will provide a gradually sounder basis for utility planning, in which fluidized-bed systems will probably have a role to play, although probably not until the 1990s.

Energy use and supply is now, like many other social processes, undergoing a profound reassessment of decision-making. Questions have arisen anew as to who pays, who benefits and who decides. This reassessment is affecting energy technology just as it affects all other sectors of the economy. The general confusion about energy decisions is compounded by the physical scale, the time scale and the financial scale of many decisions now under consideration. The problem is common to all areas of innovation, fluidized-bed energy technology included. Innovative decision-making is bedeviled by existing financial commitments, physical plant and specialized knowledge whose influence impedes changes in the direction of technical development. Technology no longer starts from scratch. Growing doubt about the validity of earlier judgments does not encourage decision-makers toward confidence in the even less familiar. Uncertainty affects all concerned: governments, plant manufacturers, and suppliers and users of electricity and heat. A vicious circle prevails. Each participant waits for the next to make the first move and the first commitment.

However, the vicious circle which has been blocking development of fluidized-bed energy systems is breaking at last. Government and industry are beginning to recognize both the potential and the opportunity which fluidized-bed technology represents. But there remain many obstacles -- especially economic and political -- to be surmounted. To make the potential of fluidized-bed energy technology an actuality will continue to require initiative and courage.

ORGANIZATIONS PROFILED BY INFORM
STATUS OF FLUIDIZED-BED COMBUSTION RESEARCH

Company	Primary Business	Type of System	Experimental Units	Offering a commercial product
American Electric Power Company, Inc.	Electric utility	Pressurized	NA	NA
The Babcock & Wilcox Company	Boiler manufacturer	Both	Atmospheric 3 by 3 foot, atmospheric 1 by 1 foot, atmospheric 6 by 6 foot	No
Babcock & Wilcox, Limited	Boiler manufacturer	Both	40,000 pounds-per-hour atmospheric	Yes (atmospheric up to 500,000 pounds per hour)
Battelle Memorial Institute	Research and development organization	Atmospheric	Atmospheric 6 inch diameter, atmospheric 2.4 square foot, atmospheric 5.6 square foot	No
Burns and Roe, Inc.	Architect-engineering firm	Both	None	No
Combustion Engineering, Inc.	Boiler manufacturer	Both	Atmospheric 2 by 23 inch, atmospheric 29.5 by 25 inch	No
Combustion Power Company, Inc.	Pollution control systems manufacturer	Both	1 pressurized, 3 atmospheric	No
Combustion Systems Limited	Joint managing agent	Both	NA	Yes (through B & W Limited)
Curtiss-Wright Corporation	Turbine manufacturer	Pressurized	Pressurized (1/4 bed diameter of a 13 MWe unit)	No
AB Enkoepings Vaermeverk	Electric utility	Atmospheric	NA	NA

Company	Primary Business	Type of System	Experimental Units	Offering a commercial product
Energy Resources Company	Energy and pollution control equipment	Atmospheric	Atmospheric 2 square foot, atmospheric 6 by 6 foot	Yes (atmospheric units up to 500,000 pounds per hour)
Fluidyne Engineering Corporation	Aerospace equipment manufacturer	Atmospheric	2 atmospheric 18 by 18 inch, atmospheric 40 by 64 inch	Yes (atmospheric air heaters up to 12,800,000 Btu/hour)
Foster Wheeler Corporation	Boiler manufacturer	Atmospheric	Atmospheric 1.75 square foot, atmospheric 36 square foot	Yes (atmospheric units up to 600,000 pounds per hour)
National Coal Board	National coal industry	Both	Atmospheric 1 MW (thermal) and several smaller, pressurized 8 MW (thermal)	NA
NCB (IEA Services) Limited	Managing agent	Pressurized	NA	NA
Oak Ridge National Laboratory	National energy laboratory	Atmospheric	Atmospheric 6 square foot, atmospheric 10 square inch	NA
Pope, Evans and Robbins Incorporated	Architect-engineering firm	Atmospheric	Atmospheric 30 MWe, atmospheric 0.5 MW (thermal)	Yes (in partnership with Foster Wheeler)
Stal-Laval Turbin AB	Turbine manufacturer	Pressurized	NA	No
Stone-Platt Fluidfire Limited	Boiler manufacturer	Atmospheric	Several	Yes (atmospheric up to 60,000 pounds per hour)

NA = Not Applicable

34

Part 2

Profiles of Major Organizations

AMERICAN ELECTRIC POWER COMPANY, INC.

Summary — The American Electric Power Company, Inc.(AEP), is one of the largest investor-owned utilities in the United States. Its distribution system sprawls over seven states in the Midwest, including Ohio, Indiana, and West Virginia. These are some of the nation's richest coal states, and seventeen of the company's nineteen generating stations are coal-fired. It is thus not surprising that AEP should be interested in fluidized-bed combustion technology. The American Electric Power Service Corporation, an AEP subsidiary which does engineering and development work for the parent company, is currently doing a feasibility study for a combined-cycle plant utilizing fluidized-bed boilers along with gas turbines. The AEP Service Corporation is investigating the possibility of major investment in fluidized-bed generating capacity.

Program and Technology — Because most of AEP's generating capacity is coal-fired, the stiffening of environmental standards which has taken place over the last decade has become a major preoccupation at AEP. All AEP plants designed before 1970 have had problems with particulate emissions. To date, retrofitting of precipitators to reduce these emissions has cost AEP $800 million, with $170 million being spent for four units in West Virginia alone. AEP's coal is generally high in sulfur, leading to problems with sulfur-oxide emissions as well. AEP is planning to install sulfur-oxide scrubbers in some future plants to reduce these emissions, but has been looking for alternative methods of sulfur control which might be preferable.

AEP has long had a close working relationship with the Swedish engineering firm of Stal-Laval Turbin AB (see profile). In September 1976, the director of Stal-Laval suggested that AEP ought to look into pressurized fluidized-bed combustion. In December 1976, AEP signed an agreement with Stal-Laval and with Woodall-Duckham, a subsidiary of Babcock and Wilcox Limited (U.K.), to carry out a feasibility study of a coal-fired combined-cycle power plant using Stal-Laval gas turbines and Babcock and Wilcox

(U.K.) fluidized-bed technology. In a combined-cycle system, hot gases from a fluidized-bed combustor are fed to a generator. In this way, more electricity can be generated from a given amount of coal. The design would be for a 105 MWe steam turbine and a 65-to-70 MWe gas turbine, to be incorporated into an existing plant in Ohio. The feasibility phase of the study is now ending. The next phase will be the detailed engineering of the project, lasting twelve to fourteen months. No decision to proceed to this phase has yet, however, been made. In the event of a decision by AEP's board of directors to proceed with the project, AEP would expect to receive some financial support from the Department of Energy or other sources.

AEP does all its own engineering which, according to Tillinghast, makes the company unique among U.S. utilities. A review committee, including representatives of AEP, Stal-Laval and Babcock & Wilcox (U.K.), meets every two or three months to oversee the running of the project.

AEP, Stal-Laval and Babcock & Wilcox (U.K.) are also spending $450,000 for pilot-scale combustion tests to verify the demonstration pressurized fluidized-bed combustor design at the Coal Research Establishment laboratories in Leatherhead, England.

Goals
and
Opinions

John Tillinghast, AEP Vice Chairman for Engineering and Construction, believes that the pressurized fluidized-bed combustion process AEP is investigating offers three advantages over conventional boilers. First, sulfur is removed from the coal in the combustion process, with a sorbent whose disposal causes few problems. Second, the use of combined cycles permits greater efficiency in conversion of coal energy to electricity. Third, the size and cost of boilers may be radically reduced, since at an operating pressure of 16 atmospheres everything is more compact. The small size of the units will make it possible to build them in a workshop instead of in the field and allow them to be built more quickly. AEP does not anticipate that spent sorbent will cause any significant increase in the usual coal-burning waste-disposal problem.

AEP is working with Stal-Laval and Woodall-Duckham, a subsidiary of Babcock and Wilcox (U.K.), on the pressurized fluidized-bed project. Tillinghast stated that it has been AEP's policy in the past to work with people and organizations that it knows and trusts and to rely on that trust as opposed to careful contracts. AEP will of course, want some sort of warranty on the proposed plant, but any warranty will be understood flexibly, since the project is research and development.

Tillinghast sees AEP policy as aggressive, exhibiting technological leadership. He says the company's general outlook is friendly toward good ideas and that this has in the past oriented AEP toward European technology. AEP is also cooperating with British Columbia Hydro in Canada, which has expressed an interest in combined-cycle fluidized-bed technology similar to that being developed for AEP.

Tillinghast feels that AEP does not have the time for a lot of paper studies before putting the technology in the field. AEP will look at fluidized-bed combustion carefully, attempt to understand it, and try to play an active, if not a leading, role in its development.

Address	American Electric Power Company, Inc. 2 Broadway, New York, New York 10004		

Financial Data	(figures in thousands)	1977	1976
	Sales	$2,031,247	$1,843,840
	Net income	236,894	236,984

THE BABCOCK & WILCOX COMPANY (UNITED STATES)

Summary

The Babcock & Wilcox Company is a major United States manufacturer of steam-generating equipment, tubular and refractory products, nuclear energy equipment and automated machinery. Its corporate headquarters are in New York. Sales in 1977 were $1.8 billion. Babcock & Wilcox is affiliated with Babcock & Wilcox Limited (United Kingdom), but divested itself of financial interest in the British company in 1975. Babcock & Wilcox has built several fluidized-bed test rigs and is investigating commercial prospects for industrial- and utility-size units.

Program
and
Technology

In an effort to remain in the forefront of technological development in steam-producing power-generation equipment, Babcock & Wilcox began to accelerate investigations of fluidized-bed combustion technology in 1975. Various divisions are involved in this effort.

Babcock & Wilcox has three small operating atmospheric fluidized beds at its Research Center (Alliance, Ohio) which represent a $4 million investment by Babcock & Wilcox and the Electric Power Research Institute (EPRI). (EPRI is the research and development organization of the utility industry.) The first, a three-by-three-foot unit, was built in 1975 with exploratory funds which the company has set aside for novel technologies. This unit has been used for the company's own research and development projects and for work done under contract with EPRI on aspects of sorbent utilization for sulfur-dioxide removal. The second unit, a one-by-one-foot system built in 1977, is being used to study fuel and sorbent characteristics. The third bed, six-by-six-feet, now undergoing start-up, was built under a contract with EPRI to provide information on adapting atmospheric fluidized-bed combustion technology for utility use. Babcock & Wilcox has compiled a review of fluidized-bed combustion literature for EPRI, and has made recommendations to EPRI on necessary research and development work to produce prototype and commercial designs of fluidized-bed utility boilers.

The Industrial and Marine Division of Babcock & Wilcox (North Canton, Ohio) is presently designing a product line of industrial atmospheric fluidized-bed boilers based on engineering studies and data derived from the three test units. The boilers will be capable of producing steam in the range of 50,000 to 300,000 pounds per hour. The company is still awaiting its first order.

The Fossil Power Generation Division of Babcock & Wilcox (Barberton, Ohio) is working on several designs for utility-size fluidized-bed systems. Together with Burns and Roe, Inc. and United Technologies, this division is designing a 583 MWe combined-cycle pressurized fluidized-bed power plant under a contract with the U.S. Department of Energy (DOE). Burns and Roe is the architect engineer on the project, the Pratt & Whitney Group of United Technologies is designing the gas turbine and Babcock & Wilcox is designing the novel part of the power plant, the pressurized fluidized-bed combustor. Under another DOE contract, Babcock & Wilcox and Stone and Webster, Inc. have completed a conceptual design for a 570-MWe atmospheric fluidized-bed power plant. The design calls for five main beds, each 71 by 24 feet, each divided into five compartments. Particulates will be removed by an electrostatic precipitator and then fed to a carbon burn-up cell. Babcock & Wilcox has also developed a conceptual design of a 200 MWe demonstration-size utility fluidized-bed boiler in a program jointly sponsored by the Tennessee Valley Authority and DOE. The design study included considerations for a scale-up to 800 MWe.

Goals and Opinions

Babcock & Wilcox is in the business of providing economic steam-producing power-generating equipment that will meet environmental standards. John Rackley, Manager of Babcock & Wilcox's Energy Systems Laboratory (Alliance, Ohio), says that the goal of the company's fluidized-bed combustion research and development program is to acquire enough information, experience, and in-house expertise in fluidized-bed combustion technology to compare its cost with that of a conventional coal-fired plant that uses a scrubber. Babcock & Wilcox is not advocating or promoting any specific coal-burning technology. The company merely wishes to be able to offer the appropriate technology to suit the individual customer. "You can't shove a technology down a customer's throat," says Rackley.

Rackley feels that if the government really considers fluidized-bed combustion a worthwhile technology for utilities,
it should make a commitment to fund research and development efforts and large-scale demonstration plants. According to Rackley, because of the expense and resources necessary, a "monumental effort" is required to take a novel
technology through all the stages prior to commercial offering. He says that, at present, no single corporation is
willing to risk this effort on fluidized-bed combustion.

Sandy Potterton, chief engineer for Babcock & Wilcox's
Industrial and Marine Division (North Canton, Ohio), feels
that the outlook for industrial fluidized-bed units is more
promising. The primary alternative is a conventional coal
boiler that uses a scrubber. He feels industrial customers
see a scrubber as a complicated chemical plant in itself
and regard a fluidized-bed unit as simpler to operate. He
points out, however, that before fluidized-bed combustion
can become commercially available to industries, several
technological hurdles must be overcome. The ability to
control rapid changes in output must be demonstrated. The
systems must be designed to use less limestone for sulfur
control to meet the same sulfur-dioxide emission standards
(scrubbers currently use less limestone than fluidized-bed
units). Improvements must also be made in the now inade-
quate coal- and limestone-feeding techniques.

Walt Hansen, the program manager (Fossil Power Generation Division, Barberton, Ohio) responsible for fluidized-
bed utility boiler development, pointed out that the application of fluidized-bed technology to utility boilers is more
complex than its application to industrial boilers. Further,
he believes that additional engineering and laboratory development, additional pilot-scale installations and at least one
large-scale (approximately 200 MWe) demonstration unit
which would operate at current commercial utility steam
conditions, are required to establish the total economic and
commercial potential of this technology.

| Address | The Babcock & Wilcox Company |
| | 141 East 42nd Street, New York, New York 10017 |

Financial	(figures in thousands)	1977	1976
Data	Sales	$1,877,200	$1,691,800
	Net income	61,800	53,100

BABCOCK & WILCOX LIMITED (UNITED KINGDOM)

Summary   Babcock & Wilcox Limited (B & W) has been for many years
a major British manufacturer of boilers for industrial and
other applications. In 1975, B & W set up a new company,
Babcock Product Engineering Limited (BPEL), to exploit
technical innovations including fluidized-bed combustion.
Another B & W subsidiary, Woodall-Duckham Limited, has
now joined BPEL to share B & W's fluidized-bed combustion
research and marketing activities. Since 1975, B & W has
operated what was for two years the world's largest fluidized-bed boiler. B & W is now offering fluidized-bed boilers in a range of sizes, on commercial terms and with a
standard warranty on parts and performance.

Program
and
Technology

British fluidized-bed combustion research was carried out
initially on bench-scale test rigs at the National Coal Board
laboratory in Stoke Orchard and the British Coal Utilization
Research Association laboratory in Leatherhead (see National Coal Board profile). However, these laboratories
lacked the funds necessary to move to a large-scale unit.
After the establishment of Combustion Systems Limited
(see profile), CSL and B & W agreed to collaborate on conversion of a standard boiler at B & W's manufacturing plant
at Renfrew, Scotland to atmospheric fluidized-bed coal-firing. CSL provided basic data and B & W the mechanical
design. The Renfrew fluidized-bed boiler has been supply-
ing 40,000 pounds per hour of steam to the works since
August 1975. Although it is a working boiler serving the
plant, it has also been used in a variety of ways for test
purposes, burning different coals under different emission
control regimens. Until the start-up of the Rivesville
plant (see Pope, Evans and Robbins Incorporated profile),
the Renfrew boiler was the largest fluidized-bed combustion
unit in the world. It is certainly the most completely successful. According to Frank Hart and Ted McKenzie of
Woodall-Duckham, the Renfrew boiler can trap virtually 100
percent of the sulfur from high sulfur Ohio-grade United
Kingdom coal. It uses a limestone sorbent, at a ratio of
2.5-4 to 1. Lower ratios -- requiring less limestone --
can, however, trap enough sulfur from most coals to meet
emission standards currently in effect anywhere.

B & W is also developing pressurized fluidized-bed systems. It is collaborating with Stal-Laval Limited on a utility-scale (200 MWe) pressurized fluidized-bed unit proposed for the American Electric Power Company (see profiles on Stal-Laval and AEP). A similar unit is proposed for British Columbia Hydro, Canada. B & W is leading the latter study and Stal-Laval hopes to be involved. An initial report has been submitted to B.C. Hydro, whose chairman was quoted in February, 1978, as saying that the utility was definitely interested in fluidized-bed technology.

Goals and Opinions

B & W's long-term goal is to manufacture and sell fluidized-bed boilers of all sizes to all potential users. B & W is already offering atmospheric fluidized-bed combustion boilers with a warranty in sizes up to 500,000 pounds per hour. B & W is prepared to guarantee the performance and output of the boiler and offer the usual 12 month mechanical warranty; they are also prepared to guarantee environmental emission-control performance. Frank Hart emphasizes that B & W takes a cautious approach in design: "We don't want to risk a crashing failure." He points out that the fluidized bed itself is rarely the cause of trouble, but that "it can be let down by what goes on around the edges."

B & W has used CSL technical data under license. At present, B & W expects to add a number of sub-licensees in the United States in 1978; discussions are now taking place in Ohio and elsewhere. Babcock Contractors, an 80 percent-owned affiliate of Woodall-Duckham USA (a subsidiary of B & W), is continuing negotiations with the state of Ohio on three fluidized-bed boiler contracts. One is a retrofit of a 6,000 pound-per-hour boiler at a state mental hospital and another a new 100,000 pound-per-hour boiler at a state penitentiary. The third is a 355,000 pound-per-hour boiler for the Columbus and Southern Ohio Electric Company. As of April 1978, the three contracts had encountered a hold-up in funding, but the state governor and his team are working to sort this out. Price estimates are now being submitted, and Hart is confident of a satisfactory outcome. Other projects in the United States are also under discussion, including units of up to 500,000 pounds per hour on the West Coast.

Outside the United States, B & W is negotiating with potential utility and industrial users in the Netherlands on two separate fluidized-bed projects. Hart expects to proceed to hardware in 1978: "I'll be disappointed if we don't." In Ireland a customer is considering a 40 MW unit to burn high ash coal. In South Africa and in Denmark, B & W has also found customers interested in industrial units. In countries where B & W does not own a subsidiary, it will license the fluidized-bed technology to boilermakers.

In Britain, B & W has already sold to Imperial Chemical Industries an 80,000 pound-per-hour boiler which is presently burning oil, but which is likely to be modified to fluidized-bed coal-firing. Hart points out that with Britain's reserves of North Sea oil, "no one will burn coal until he has to." He says there is nevertheless general recognition that a full-scale British demonstration plant for fluidized-bed combustion is needed, "if only to help sell the technology overseas." Government backing is available, "but we have to decide how to tap it."

Hart considers that industrial users will choose atmospheric fluidized-bed units. However, he feels utilities will choose pressurized fluidized-bed units, especially to achieve higher cycle efficiency through combined-cycle operation. Hart stresses that the larger sizes of atmospheric fluidized-bed systems do have advantages. He says that the output of a small atmospheric fluidized-bed unit cannot easily be varied to match demand. "It's easier with a bigger unit." B & W (U.K.) is prepared to design and build fluidized-bed units, both atmospheric and pressurized, to whatever specifications the customer requires. The company clearly expects to be very busy doing so.

Address	Babcock & Wilcox, Limited St. James's Square, London, England SW1Y 4LN

Financial Data	(figures in thousands)	1976
	Sales	Ł629,932
	Net income	20,157

BATTELLE MEMORIAL INSTITUTE

Summary One of the world's largest research and development organizations, with an annual research budget of $184 million, Battelle Memorial Institute has been engaged in scientific and technical research since the late 1920s. The Institute carries out a wide range of research projects under contracts with government and industry, nationally and internationally, as well as a limited number of its own projects. Its headquarters are at the Battelle Columbus Laboratories, Columbus, Ohio.

In the late 1930s and 1940s, Battelle was heavily involved in research on advanced methods of using coal. Though the Institute de-emphasized this research during the following oil-dominated decades, many of these experienced coal people remained with Battelle. In 1968 and 1969, Battelle staff anticipated that cheap oil would not last indefinitely. In 1973, Sherwood Fawcett, President of Battelle, announced the "Battelle Energy Program" to develop alternative energy technologies based primarily on coal, specifically Ohio coal. The Institute wanted to work in an area in which the federal government was not already interested. Thus, Battelle stayed out of coal gasification and liquefaction. Instead, Battelle came up with a unique approach to fluidized-bed combustion, the "multi-solid fluidized bed," which can produce four times as much heat as a conventional fluidized bed of equivalent size. Since 1974, Battelle has spent $800,000 on fluidized-bed combustion work, out of a Battelle Energy Program annual budget of $3 to $4 million. There are at present fourteen scientists and engineers at Battelle's Columbus Laboratories working full time on the fluidized-bed project.

Program
and
Technology

Because of research work done for the chemical and oil industries, Battelle has had extensive experience with fluidized-bed technology. After surveying the fluidized-bed research and development programs within the United States, Battelle decided that these programs were very similar and that there was room for a truly innovative approach to the fluidized-bed combustion of coal. Called the Battelle multi-solid fluidized-bed combustion process (MSFBC), this unique approach is able to produce heat four times faster than a conventional system because it overcomes an inherent limitation in conventional fluidized-bed combustion. The limitation arises as follows. The rate at which coal can be burned in a given area of bed depends on how fast combustion air can be provided. However, if air is blown in faster, more particles are lifted out of the bed and tend to be carried away before they are fully burned. There is thus a trade-off: if coal is added more rapidly, air must be added more rapidly in order to burn it; but the increased addition of air means that the extra coal is not fully burned, defeating the purpose of the exercise. There is thus a practical upper limit on the rate of heat release per unit area of bed.

To overcome this problem, the Battelle designers came up with what they christened a "multi-solid" bed. The Battelle system uses a bed of the inert and dense mineral hematite, in the form of fine sand. The weight of the hematite particles means that fluidizing air can be blown up through this bed much faster, without lifting the particles above the surface. When coal particles are fed into the bed, they burn; the coal ash, being much less dense than the hematite, is lifted rapidly out of the bed in a rising column of hot gas and dust. Boiler tubes may be placed in this column. The hot gas and dust are then fed out the top of the fluidized bed and across to an adjoining hopper into which the hot dust falls, while the combustion gases pass onward out of the system. The hopper of hot dust can also be fluidized, but at a lower air velocity. Boiler tubes may likewise be placed in this second fluidized bed, in which no combustion is taking place but which is at a temperature close to that of the combustion bed alongside. The dust from the hopper can be fed back into the combustion fluidized bed at a controlled rate, to ensure complete burn up of the fuel and to allow a sophisticated control of the system's operating temperature.

Battelle found enough money in 1974 to build a bench-scale test MSFBC rig with a six-inch diameter combustor. Then, in December 1975, the U.S. Energy Research and Development Administration (ERDA, now the U.S. Department of Energy [DOE]) put out a "Program Opportunity Notice" for industrial demonstration fluidized-bed systems. The notice required the prospective contractor to put up part of the money for the project ("cost sharing"). Battelle did not have sufficient funds for such an arrangement. Battelle also considered it advisable to be able to call on a boiler manufacturer's expertise in the design and construction of the demonstration facility. Battelle approached every major boiler manufacturer about the proposal, but only the Foster Wheeler Corporation responded with enthusiasm and financial backing.

Battelle and Foster Wheeler agreed to collaborate. Together they landed an ERDA/DOE contract for $5.2 million to develop the MSFBC concept. The Department of Energy paid the cost of building a 2.4 square-foot test rig producing 4,000 pounds per hour of steam. This unit is presently being commissioned at the Battelle Columbus Laboratories. It will provide data for a 25,000 pound-per-hour unit, costing $3.6 million, which is also being built at Battelle Columbus Laboratories. The Department of Energy will pay half the cost of this latter unit, and Battelle and Foster Wheeler will pay the other half, along with a $250,000 contribution from the Ohio Department of Energy.

Battelle says that the MSFBC system achieves 95 percent burn-up of carbon. It can remove 83.85 percent, perhaps up to 90 percent, of the sulfur from Ohio coal containing 4 percent sulfur. The spent limestone is sent to Westinghouse Electric Corporation for study of possible by-product uses. Sorbent disposal "doesn't look threatening," according to Carl Lyons, Battelle's Associate Director for Project Management and Program Development.

The technical problem of most concern to the Battelle team is the possible erosion of boiler tubes. Like other fluidized-bed investigators, Battelle is concerned with coal-handling difficulties, expecially the feeding of wet coal.

Battelle has a $1 million contract from the U.S. Environmental Protection Agency (EPA) to evaluate the environmental impact of fluidized-bed combustion.

Goals and Opinions

Battelle's goal is to commercialize its MSFBC system rapidly. The state of Ohio has given the project a boost. During the energy shortage in the Midwest in 1976, Ohio Governor

James Rhodes asked Battelle President Fawcett what technology might be available to use coal cleanly. The two main immediate possibilities appeared to be scrubbers and fluidized-bed systems. The Governor formed a committee which considered these and several other technologies and selected fluidized-bed combustion. The Governor visited Battelle, saw the original small MSFBC rig and was sufficiently impressed to "stir the pot" in the media and elsewhere. However, the Governor was concerned about the probable time lag before MSFBC systems would be available commercially. Battelle suggested to the Governor that, if the state were to take responsibility for offering a warranty on the system, it would ease the path to commercialization. In the meantime, Babcock & Wilcox (U.K.) came to Ohio and, in Lyons' words, "did a great selling job. They could build a fluidized-boiler and offer a warranty immediately. The Governor's committee went with Babcock & Wilcox." (For further details, see profile on Babcock & Wilcox [U.K.].) The Ohio Department of Energy considers the MSFBC design a "second-generation" fluidized-bed system. Battelle has estimated the cost of retrofitting Battelle's Columbus Laboratories' Number Three boiler and another, on a nearby site. Both these projects, if undertaken, would involve the sharing of costs between Battelle and the state of Ohio.

Battelle organized a conference in September 1977, at the instigation of Governor Rhodes, to help to acquaint industrial and utility customers with the fluidized-bed concept. As a result, there have been over a dozen inquiries from textile, paper and other companies interested in possible fluidized-bed applications.

Lyons says that he was "very surprised" to realize that Battelle's MSFBC system was "the only unique approach" to fluidized-bed combustion. Battelle is also looking into pressurized fluidized-bed technology, thus far only at the thinking stage.

Address   Battelle Memorial Institute
           505 King Avenue, Columbus, Ohio 43201

Financial (figures in thousands)   1977   1976
Data   Research volume $224,400 $184,800

BURNS AND ROE, INC.

Summary

Burns and Roe, Inc. is an architect-engineering firm with considerable experience in the design of energy systems. Much of the company's work has been in the engineering, design and construction of conventional fossil and nuclear power generating stations for utilities. Burns and Roe is presently at work on two conceptual designs of fluidized-bed combustion plants.

Program
and
Technology

In examining the systems to which utilities might turn for power generation in the future, Burns and Roe decided that fluidized-bed combustion showed great promise and that the company should develop its capability in this area. This decision led, in 1976, to two separate contracts with the U.S. Energy Research and Development Administration (ERDA, now the Department of Energy, DOE).

One contract involves the conceptual design of a 570 MWe atmospheric fluidized-bed plant. The other contract, in association with the Babcock & Wilcox Company (U.S.) and United Technologies, Inc., is for the design of a 583 MWe combined-cycle pressurized fluidized-bed unit.

Upon the award of the contract for the atmospheric fluidized-bed plant, Burns and Roe prepared and issued design specifications for the project to several boiler manufacturers and, on the basis of the responses, awarded a subcontract to Combustion Engineering, Inc. The contract specifies that the design be for a power plant to burn Western coal with a .9-3.25 percent sulfur content and a 20 percent moisture content. The boiler design calls for twenty 9 by 36 foot beds, all on a single level. DOE required that the plant be designed for a 20 year life with 95 percent availability.

The conceptual design and economic and environmental analysis of the plant are complete and the design has been submitted to DOE for review and comment. Burns and Roe's economic analysis shows that the atmospheric fluidized-bed plant has the potential to be less costly than a conventional pulverized coal installation using a scrubber system. DOE has proposed additional work in an attempt to improve the cost and performance of the initial design. Burns and Roe's new mandate is to redesign the plant to meet a 90 percent sulfur removal standard, and to lower the calcium/sulfur ratio and improve the combustion efficiency of the system (both primary concerns for utilities).

The design for the pressurized fluidized-bed plant calls for the combustors to exhaust to gas turbines followed by a steam-cooled atmospheric fluidized-bed boiler. Thus, this design is fundamentally different from the other DOE-funded combined-cycle pressurized fluidized-bed designs of the Curtiss-Wright Corporation and the General Electric Company (see profile and appendix). As data for the design, Burns and Roe used results from testing done by the Babcock & Wilcox Company (U.S.), Babcock & Wilcox, Limited (U.K.) and DOE-funded projects. According to Burns and Roe, their combined-cycle design offers a thermodynamic efficiency of 40 percent, approximately 5 percent higher than a conventional plant. The design work is complete, and a final draft of the design is being readied for submittal to DOE. Donald Huber, Burns and Roe's project manager for the design, feels that the project demonstrates that combined-cycle pressurized fluidized-bed units can prove economic, if the necessary development work can be successfully completed.

Goals
and
Opinions

According to Bill Bradley, project engineer for Burns and Roe's atmospheric fluidized-bed design, his objective is to determine whether a plant can be designed to meet new U.S. Environmental Protection Agency sulfur-dioxide emission standards and to compete economically with a pulverized coal boiler using a scrubber. Bradley sees fluidized-bed combustion as a potential near-term solution to the use of scrubbers; a bridge to the next century when coal gasification, coal liquefaction, magneto-hydrodynamics, fusion and solar energy may become commercial realities. Bradley feels the sulfur-capturing ability of fluidized-bed combustion systems has been demonstrated and that the technology has the potential to meet even stricter regulations. What is needed now, according to Bradley, is a utility-sized demonstration plant, preferably more than one, that can utilize a variety of fuels and prove control of rapid changes of output.

Bradley says that until these capabilities are demonstrated, utilities will not acquire confidence in the technology's ability, and he feels the government must assist in the funding of fluidized-bed demonstration projects until the technology is successfully commercialized. He thinks the government could also aid in fluidized-bed combustion's development by creating a stabilized sulfur-dioxide emission standard, thus "giving us a stationary target to shoot at."

Donald Huber says that the ultimate goal of the pressurized fluidized-bed design is to prove that combined-cycle pressurized fluidized-bed systems are economic. He feels that the project has proved this and it is now a question of getting it down on paper. Burns and Roe hopes to gain sufficient knowledge and experience from this design work to apply it to design and engineering activities for utilities. Huber believes pressurized fluidized-bed units will be commercially available in the late 1980s. The most significant present problems, according to Huber, are in the areas of materials and fuel feed. A larger problem is the inherent reluctance of utilities to invest in any technology that has not been demonstrated. He concurs with Bill Bradley and would like to see the government fund one or more demonstration plants.

Address   Burns and Roe, Inc.
               550 Kinderkamack Road, Oradell, New Jersey   07649

Financial   Burns and Roe is privately owned; no financial data is
Data   available.

COMBUSTION ENGINEERING, INC.

Summary   Combustion Engineering, Inc. is one of the largest manufacturers of steam-generating systems and equipment in the world. The company also designs, engineers and constructs chemical process plants, petroleum refineries and facilities for the petrochemical, metallurgical, and pulp and paper industries. Combustion Engineering's corporate headquarters are in Stamford, Connecticut and its 1978 sales were over $2 billion. The Power Systems Group of Combustion Engineering is engaged in four fluidized-bed combustion projects.

Program
and
Technology

In 1976, the company began a study of the feasibility of retrofitting Consolidated Edison's 500 MWe Arthur Kill #3 unit on Staten Island, a borough of New York City, with an atmospheric fluidized-bed boiler. The New York State Energy Research and Development Authority (NYSERDA) and the Electric Power Research Institute (EPRI) funded the work. In the course of the study, Combustion Engineering was to do a conceptual design of an atmospheric fluidized-bed retrofit, develop a cost comparison between the fluidized-bed combustor and the present system with a scrubber added, and identify the potential problem areas and areas requiring further development in fluidized-bed combustion. The study concluded that it was technically feasible to make the retrofit and that the cost was comparable to that of a conventionally fired system using a scrubber. However, according to a NYSERDA spokesman, "The option is not extremely attractive economically. The economics look better for industrial-size retrofits."

In 1977, Combustion Engineering was one of three boiler manufacturers (the other two were Babcock & Wilcox (U.S.) and Foster Wheeler Corporation) awarded a contract by the Tennessee Valley Authority (TVA) to do a conceptual design for an 800 MWe demonstration plant. Combustion Engineering's preliminary design for the contract was based on the same design parameters as those utilized in the Pope, Evans and Robbins Incorporated Rivesville plant (see Pope, Evans and Robbins profile).

The company is working on a third design for a 570 MWe atmospheric fluidized-bed steam generator to burn Western coal. Combustion Engineering is a subcontractor to Burns and Roe, Inc., which has the contract with the Department of Energy (DOE). For the project, Combustion Engineering is modifying its TVA design so that the system will be able to burn the DOE-specified Western coal, which has a 7.8 percent ash, a 28 percent moisture and a .9 to 3.25 percent sulfur content. A primary feature of this design is that the 20 individual fluidized-bed cells which comprise the unit are not stacked one on top of the other but are all on the same level, "ranch style." Combustion Engineering feels this feature will allow for easier maintenance and operation.

The fourth project is part of DOE's demonstration program for industrial applications of fluidized-bed combustion processes. The project is to design and construct a 50,000 pound-per-hour boiler for the U.S. Navy Great Lakes Training Station in Illinois. The total funding for the project is $8.316 million, with Combustion Engineering providing $1.061 million, the U.S. Navy at Great Lakes $637,000, the Department of Energy $5.868 million, and the State of Illinois (through its Coal Development Bond Funds program) $750,000. The project was initiated on July 30, 1976 and the plant is scheduled to be in operation before the end of 1980.

The goal of the project is to develop a commercial atmospheric fluidized-bed "package" boiler; that is, a boiler that is capable of being assembled in the shop and shipped by rail. There are two phases to the project. The first phase, carried out by Combustion Engineering's Kreisinger Development Laboratory in Windsor, Connecticut, consisted of designing and constructing a three-by-three foot fluidized-bed test unit for investigating fluidized-bed combustion characteristics and assessing environmental impacts. This first-phase unit is presently operating. The second phase will be the detail design, construction, and testing of the U.S. Navy Great Lakes Training Station boiler.

Goals and Opinions

Although Combustion Engineering provided INFORM with extensive written material on its fluidized-bed combustion program, the company declined to state its goals or opinions.

Address   Combustion Engineering, Inc.
900 Long Ridge Road, Stamford, Connecticut 06902

Financial Data	(figures in thousands)	1977	1976
	Sales	$2,044,764	$1,830,925
	Net income	67,189	54,203

COMBUSTION POWER COMPANY, INC.

Summary

Combustion Power Company, Inc., is a Menlo Park, California development firm specializing in environmental systems. Since 1975, it has been a wholly owned subsidiary of the Weyerhauser Company, which had sales of $2.86 billion in 1976. Combustion Power's interest in fluidized-bed combustion grew out of work it did on burning municipal wastes. Combustion Power is offering its fluidized-bed combustion units commercially, and has sold two units to Weyerhauser.

Program and Technology

Combustion Power conceived the CPU-400, a pressurized fluidized-bed combustor coupled to a gas turbine, in 1967, to burn municipal wastes. With funds from a U.S. Environmental Protection Agency (EPA) contract, Combustion Power designed and constructed a pilot plant. The plant began operating in 1973, consuming up to 100 tons of garbage daily to produce 1 MW of electricity. Combustion Power designed the CPU-400 to burn any low-grade fuel, including high-sulfur coal and wood wastes. The Energy Research and Development Administration (ERDA, now the Department of Energy [DOE]), began funding the project in 1973. The ERDA contract, for $4 million, was to develop the capability of burning high-sulfur coal.

In the CPU-400 process, coal is crushed, mixed with dolomite, and fed into a pressurized fluidized-bed combustor. The hot gas from the combustor passes through a three-stage separator to remove particulates which would otherwise corrode the blades of the gas turbine.

Combustion Power tested several different coals at varying bed temperatures, fluidizing velocities and calcium/sulfur ration. The major problem encountered in the testing was that the separator system did not remove enough of the particulates to protect the turbine blades. Combustion Power completed experimental work under the ERDA contract in 1976.

The company is now offering fluidized-bed combustion systems for energy recovery from waste materials commercially. Weyerhauser bought two units, a 22 foot internal Combustion Power tested several different coals at varying bed temperatures, fluidizing velocities, and calcium/sulfur ratios. The major problem encountered in the testing was percent dirt, sand and rock. One unit provides hot gas to a boiler, the other supplies hot gas to a dryer.

Combustion Power is presently operating four test units. One of these is pressurized; the other three are atmospheric. The company's current research and development effort is focused on the troublesome hot gas clean-up problem.

Goals and Opinions	Combustion Power's long-term goal is to sell its fluidized-bed combustion systems to industries and utilities. Dale Moody, a Combustion Power marketing manager, feels customers still hesitate to buy a fluidized-bed combustor because it is not yet a proven product. He feels general adoption of fluidized-bed combustion systems will take time due to the high capital costs involved in fuel- and sorbent-handling equipment along with a need to prove the technology's capability to burn a wider variety of fuels. Moody thinks that it is a misconception to see fluidized-bed units as a replacement for all other coal-burning technologies. Fluidized-bed units cannot, he says, compete economically in geographical areas where pulverized-coal units can burn low-sulfur coal. Only in areas where low-sulfur coal is not available, or where other specific environmental factors come into play, will fluidized-bed systems be used. For example, because a fluidized-bed combustor operates at a lower temperature than a pulverized-coal unit, the fluidized-bed unit has lower nitrogen-oxide emissions. This is a definite advantage in Combustion Power's home state of California, where nitrogen-oxide emission standards are much stricter than the federal standards.
Address	Combustion Power Company, Inc. 1346 Willow Road, Menlo Park, California 94025
Financial Data	Combustion Power Company was bought by Weyerhauser in 1975; sales and income figures are no longer divulged.

COMBUSTION SYSTEMS LIMITED

Summary

Combustion Systems Limited (CSL) was set up in London in 1972 by its owners, three British organizations with an interest in fluidized-bed combustion. The three are: The National Coal Board (NCB), which owns and operates Britain's coal industry (see profile); British Petroleum (BP), a major multi-national oil company; and the National Research Development Corporation, a government-owned agency which provides capital and management support for technological innovation. CSL is the joint "managing agent" for the involvement of the three businesses in fluidized-bed development, serving as an information clearing house and a marketing and licensing agent. CSL sponsors program and technology research work at the two Coal Research Establishment laboratories of the NCB and at the Sunbury laboratory of BP. All three CSL partners have access to results. CSL also has an agreement with Woodall-Duckham, a subsidiary of Babcock & Wilcox Limited (B & W [U.K.]), for a joint operation of the large 40,000 pound-per-hour atmospheric fluidized-bed boiler at the B & W plant in Renfrew, Scotland. (See NCB and B & W [U.K.] profiles.)

Program
and
Technology

CSL does not itself build fluidized-bed hardware. Rather, it draws together the information generated by its partners and acquired through its link with Woodall-Duckham. It synthesizes the various engineering and economic data for commercial purposes. For example, the results of the River Don conversion (see NCB profile) by the NCB will be available through CSL. CSL in turn, through its field agents and technical people, can prompt research by its partners on appropriate test rigs. Such research is always carried out with a commercial objective in view.

According to CSL managing director David Knights, CSL is now receiving a steady stream of inquiries on fluidized-bed combustion from prospective customers. Recent inquiries have come from the United Kingdom, United States, Japan, Denmark, the Netherlands, and West Germany, and from electrical utilities, industries, and boilermakers.

CSL is offering licenses to use its owners' technology on a non-exclusive basis. It also has a licensing agreement with Woodall-Duckham and is thus indirectly involved in present contractual arrangements with the state of Ohio (see B & W [U.K.] profile). CSL recently signed its first license agreement with a non-British firm, Johnston Boiler Company of Michigan, for small, packaged, industrial atmospheric fluidized-bed boilers. Knights anticipates that further agreements will be signed within the coming six months. He says CSL has no problem gaining access to the United States market. He feels that CSL is significantly ahead of its competition as the result of its considerable experience at its owners' laboratories and of its joint operation of the Renfrew fluidized-bed unit for more than two years.

Goals and Opinions

Knights considers that atmospheric fluidized-bed systems are "ideal for industrial applications and even power generation within industry, but not for main utilities." He believes that "pressurized fluidized-bed systems will be the choice for main utility systems."

Address

Combustion Systems Limited
195 Knightsbridge, London, England   SW7 1RD

Financial Data

No financial data available.

CURTISS-WRIGHT CORPORATION

Summary
The Curtiss-Wright Corporation is an engineering firm whose origins date back to the Wright brothers, who gave the firm its name. Curtiss-Wright, with sales of over $350 million in 1976, is now one of the world's major manufacturers of turbines, both as jet engines and as gas turbine-driven electric-power generation systems. Curtiss-Wright also owns 65 percent of Dorr-Oliver Incorporated, which has designed, built and sold over 600 fluidized-bed units for ore roasting, waste incineration and other industrial applications. Together with Dorr-Oliver, Curtiss-Wright is currently working to develop a system coupling fluidized-bed combustion systems and gas turbines.

Program and Technology
The turbine expertise of Curtiss-Wright and the fluidized-bed expertise of Dorr-Oliver led, in January 1976, to an Energy Research and Development Administration (ERDA, now U.S. Department of Energy [DOE]) contract to design, build and operate a 13 MWe coal-fired gas-turbine system using a pressurized fluidized bed. Some 100 scientists and engineers at Curtiss-Wright are now at work on the project. The plant is under construction at Curtiss-Wright's main headquarters in Wood-Ridge, New Jersey. Curtiss-Wright is contributing an additional one-third to ERDA/DOE's $30.3 million. Major sub-contractors are Dorr-Oliver and Stone & Webster, Inc. An existing natural gas-fueled gas turbine and a waste-heat recovery boiler which provide electric power and steam for processing and heating purposes at Wood-Ridge will be used in the program. In addition, coal-handling equipment and service installations, part of an existing boiler house which was originally coal-fired, are being taken out of mothballs and recommissioned. The new pressurized fluidized-bed combustor will be coupled to an existing gas turbine and electricity generator. Hot gases and hot air from the fluidized bed will drive the blades of the gas turbine to turn a 7 MWe generator.

The fluidized-bed unit will use a bed consisting mainly of crushed dolomite. Research by the National Research and Development Corporation in the United Kingdom has shown that under pressure, dolomite traps sulfur more effectively than does limestone. The bed temperature will be held at 1650°F to optimize sulfur retention, although the gas turbine would operate more efficiently with hotter gases.

The gas-turbine compressor will supply the air for the fluidized-bed unit. One-third of the air from the compressor will serve as combustion air. The remaining two-thirds of the air from the compressor will pass through tubes inside the bed. This air will be heated to bed temperature, but will not come in contact with the coal. It will therefore remain uncontaminated by particulates and gases. The contaminated combustion gas will be cleaned. It will then rejoin the other two-thirds of the air stream, now hot. The clean hot air will dilute any remaining particulates and vapors, further reducing their impact when the whole stream is fed into the gas turbine.

Nevertheless, the turbine blades must be protected against erosion and corrosion by the combustion gases and any particulates still carried along. Such blade damage has been the main reason for failure of previous attempts by other investigators to devise a coal-fired gas turbine. Curtiss-Wright will use a special turbine blade design. It has an inner core through which air will be blown. The air will emerge through wire mesh skin, creating a thin "boundary layer" of higher-pressure air immediately outside the surface of the blade. The boundary layer will deflect both hot gases and any remaining particulates. The blade will also operate at a substantially cooler temperature than the hot gas, thus further reducing susceptibility to corrosion.

As in any gas turbine, the exhaust gases emerge still hot enough to pass through another boiler and produce steam. This steam may be fed into a steam turbine to produce electricity. Such "combined cycles," as they are called, would convert substantially more of the original coal energy into electricity -- more than 50 percent -- than is possible in a conventional power plant, which cannot usually convert more than about 40 percent. Alternatively, as will be the case for the Wood-Ridge pilot plant, the steam can itself be used for industrial processes. This again converts significantly more of the coal into useful energy.

The ERDA/DOE contract also involves producing a conceptual design for a 500 MWe full-scale combined-cycle plant. Curtiss-Wright envisages a plant consisting of three modules. Each module will produce 165 MWe, with all turbines operating at full output. To enable the three-module station to follow electrical demand, individual turbines can be shut down separately to reduce power.

Already in operation at Curtiss-Wright is a pressurized fluidized-bed combustion system which is one-quarter the diameter of the 13 MWe pilot plant and uses the same coal, dolomite, gas clean-up, ash-handling, combustor-internals and control systems. The purpose of the smaller system is to provide actual operating design, material, performance and control data as support to the design process now under way on the pilot plant. The operating unit was designed, built and brought on line within one year of the start of contract.

Goals and Opinions

Curtiss-Wright's long-term aim is to sell combined-cycle pressurized fluidized-bed systems to utilities. Its objective for the next two years, however, is to complete design and construction of the Wood Ridge pilot plant successfully and to design the 500 MWe full-scale plant. As of October 1977, the work was on schedule, one and one-half years into the five-year project. When the pilot plant is complete, there will be two years of test operation. They are using two different high-sulfur coals, selected by Curtiss-Wright and by DOE for test purposes.

The chief operating officers of Curtiss-Wright and Dorr-Oliver made the original decision to involve the company in pressurized fluidized-bed development. They felt that it looked like a promising business proposition and that work on pressurized fluidized-bed technology would sharpen skills and provide experience for other tasks. However, Curtiss-Wright would not have embarked on a project of pilot-plant magnitude without DOE money.

Curtiss-Wright plans to petition for special patent privileges in application areas outside the specific subject of the DOE contract. It says that such a procedure is standard. Arnold Kossar, Vice President of Curtiss-Wright, regards warranties as the "soft spot" inhibiting demonstration energy plants. He wonders how utilities can justify putting out money for a substantial plant on the basis of a pilot plant. Utilities "need their own confidence level about return on capital." For this reason, Kossar and Seymour Moskowitz, Director of Energy Systems R&D, agree that the DOE should provide cheap capital to assist in the financing of fluidized-bed projects. Moskowitz points out that the Internal Revenue Service allows bonds for pollution-control technology such as stack-gas clean-up to be tax-exempt. However, at present, if sulfur is removed in the combustion process, no such exemption applies. Kossar and Moskowitz would like to see such "institutional asymmetries" eliminated.

Curtiss-Wright is also interested in atmospheric fluidized-bed systems. Kossar and Moskowitz feel that atmospheric fluidized-bed technology will be appropriate for industrial users, with utilities being the major market for pressurized fluidized-bed technology.

Curtiss-Wright and Dorr-Oliver are actively engaged in exporting gas turbines and fluidized-bed technology worldwide. Kossar and Moskowitz consider fluidized-bed energy systems a feasible addition to their export line. They might even be willing to take more financial risk with fluidized-bed systems than they would with conventional ones. However, in the long term, Curtiss-Wright is mainly concerned with establishing a preferred position selling the next generation of gas turbines to utilities, and feels that utilities are bound to go in the direction of coal-fired combined-cycle systems using pressurized fluidized-bed combustion.

Address	Curtiss-Wright Corporation
	One Passaic Street, Wood Ridge, New Jersey 07075

Financial Data	(figures in thousands)	1977	1976
	Sales	$323,128	$357,266
	Net income	16,299	19,078

ENERGY RESOURCES COMPANY

Summary

The Energy Resources Company (ERCO), a small, privately-owned company founded in 1973, designs energy and pollution control equipment. ERCO has considerable experience in fluidized-bed technology and is presently working on both pyrolysis and combustion system designs. ERCO has sold two fluidized-bed combustion units to industrial customers.

Program
and
Technology

ERCO began purely as a consulting organization, making paper studies for the government on energy and environmental issues. In 1974, ERCO obtained a contract from the U.S. Environmental Protection Agency (EPA) to do studies on certain aspects of fluidized-bed pyrolysis technology.

ERCO decided to investigate fluidized-bed combustion in 1975 when Dr. James Porter, an engineer experienced in the technology, joined the company. After several fruitless attempts to obtain research and development money from the Energy Research and Development Administration (ERDA), ERCO decided to use company funds for tests. It converted its fluidized-bed pyrolysis unit built for EPA work into a fluidized-bed boiler. ERCO then tested sixteen different coals at fluidizing velocities from four to sixteen feet per second in the two square-foot bed-area combustor. (ERCO declined to provide INFORM with information on the structure of the test unit and the data derived from it.)

ERCO is now building a six-foot pilot unit for further tests. The company is offering to design and contract for the building of industrial-size atmospheric fluidized-bed units capable of producing up to 500,000 pounds of steam per hour.

Goals
and
Opinions

ERCO claims to have developed a workable design for industrial-size atmospheric fluidized-bed units. Porter, who is Vice President and Technical Director for ERCO's Energy Systems Division, says ERCO is at a disadvantage in seeking government research and development funds and industrial contracts. The company is small and new, and is competing against large firms with years of experience in steam production and power generation. Porter says the usual response is, "Who's ERCO?" The company will use its six-by-six foot unit to obtain visibility and credibility for its fluidized-bed program.

Porter says ERCO's primary advantages are its versatile pilot unit and the intelligence and innovative ability of its personnel. Andrew Himmelblau, Design Engineer, feels that the company has synthesized other companies' fluidized-bed work and has learned from the others' mistakes. He cites as the example the coal-feed system ERCO is now patenting. This system will have one coal-feed point for every 100 square feet of bed area, a figure much better than the one feed point per nine square feet which is now in use at the Rivesville plant (see Pope, Evans and Robbins Incorporated profile).

Himmelblau says that ERCO feels confident enough to go out and build boilers immediately. He says that obviously, systems will improve as the company gains more experience operating fluidized-bed combustors, but that ERCO's present designs are conservative enough for the boilers to operate without problems.

Porter feels that government funds for fluidized-bed combustion should be spent on basic research in recognized problem areas rather than on construction of large-scale demonstration plants. He says that if utilities feel the need for demonstration of fluidized-bed technology before purchasing fluidized-bed units, the utility industry should support a demonstration program. Porter would like to see funds allocated for research to minimize the amount of lime-stone needed to remove a certain amount of sulfur and to increase the overall energy efficiency of fluidized-bed combustion systems.

Address   Energy Resources Company
                 185 Alewife Brook Parkway, Cambridge, Massachusetts 02138

Financial   Energy Resources Company is privately owned; no financial
Data   data is available.

AB ENKOEPINGS VAERMEVERK

Summary

Enkoeping is a town of 20,000 near Stockholm, Sweden, served by a district-heating company called AB Enkoepings Vaermeverk. In 1973, the company was cut off from its fuel-oil supply for a time. As a result, the company undertook development of a fluidized-bed unit which could burn any available fuel. In April 1975, the Swedish National Environmental Protection Board gave permission for the construction of the first fluidized-bed district-heating boiler at Enkoeping. The plant has been operating successfully since spring 1978. It is the world's second-largest fluidized-bed unit, exceeded in size only by the Rivesville plant (see Pope, Evans, and Robbins profile).

Program
and
Technology

The Number Three boiler at Enkoeping is a 25 MW (thermal) atmospheric fluidized-bed boiler, producing water at $120^{\circ}C$ and ten atmospheres pressure, to supply the distribution pipes of the district-heating grid. The contractor selected for the project was Kymi, Kymmene Metal, of Heinola, Finland. The combustor was built by Mustad and Søn of Gjøvik, Norway. The installed cost of the unit (in United States dollars) was $1.2 million, of which 50 percent was financed by the Swedish government as support for the demonstration project. An additional $0.1 million has also been provided by the Swedish government to finance experimental use of different fuels, including coal, peat and wood chips. The fluidized-bed unit was first brought on stream in February 1978, burning fuel oil. The first coal firing took place in March 1978.

The system uses dolomite rather than limestone as a sulfur-sorbent when burning coal. To control particulate emissions, the unit is equipped with a novel textile filter, which is much more compact than conventional particulate filters.

Goals
and
Opinions

The choice of an atmospheric fluidized-bed unit was prompted by the possibility of flexible choice of fuel, rather than by cost, efficiency or environmental reasons. The technical-scientific project manager, David Arthursson, expects the unit to be able to remove up to 85 percent of the sulfur

from coal containing 1.5 percent sulfur. More could be
removed, but not economically. Arthursson considers that
a coal-cleaning process, perhaps one developed by the
Swedish firm Sala International, would be a better approach
to the use of higher-sulfur coal. In any case, it is expected
that the unit will also burn both peat and wood chips on a
commercial basis, both being indigenous fuels. Regard-
less of which fuel is used, Arthursson and his colleagues
expect no difficulties with disposal of solid waste.

Address   AB Enkoepings Vaermeverk
                 S-199 00 Enkoeping, Sweden

Financial   No financial data available.
Data

FLUIDYNE ENGINEERING CORPORATION

Summary

Fluidyne Engineering Corporation, located in Minneapolis, is a small, privately-owned designer and manufacturer of aerospace equipment. The company's interest in fluidized-bed combustion came about as the result of Fluidyne's desire to diversify and its feeling that it could deal well with experimental technologies. Fluidyne now has a $4 million contract to design and build an atmospheric fluidized-bed air-heater and air-distribution system for the Owatonna Tool Company of Owatonna, Minnesota.

Program
and
Technology

In 1971, interested in diversifying the company, Fluidyne's Executive Committee closely examined several experimental coal-utilization processes. Fluidized-bed combustion appeared the most promising. The committee felt that Fluidyne's background in aerodynamics and expertise in the development of experimental technologies could be profitably applied to fluidized-bed combustion. In 1973, the company built a small boiler with an eighteen-by-eighteen-inch bed area, and used it for test purposes for two years. By 1975, Fluidyne felt the fluidized-bed steam-boiler field to be increasingly crowded and switched its efforts to fluidized-bed air heaters. Its rationale was that small industries which need hot air for process purposes would be a good market for Fluidyne to pursue.

Internal research and development on fluidized-bed air heaters led, in June 1976, to a contract with the U.S. Energy Research and Development Administration (ERDA, now U.S. Department of Energy [DOE]) to design and build a system to supply high-temperature, low-pressure air for process and heating purposes at the Owatonna Tool Company. The Owatonna design calls for two five-by-eleven-foot beds capable of producing 1 to 28 million Btus per hour. The system is designed to burn a wide variety of fuels including coal, lignite, petroleum, coke, peat, wood and paper. With particulates recycled, it will have a combustion efficiency ex exceeding 95 percent. The sorbent will be dolomite, and according to Fluidyne, sulfur-oxide emissions will be well below the U.S. Environmental Protection Agency (EPA) requirements. Final design review has been completed and construction of the facility has recently commenced. Fluidyne hopes to have the unit ready for the 1978-to-1979 heating season.

In addition to the Owatonna project, Fluidyne's fluidized-bed program includes three operating fluidized beds. Two are used for fuel testing while the third, a 5 million-Btu-per-hour heater, is used for demonstrations and studies of customer systems.

Goals and Opinions

Fluidyne hopes eventually to become a major supplier of air heaters to small industrial users. Robert Sirany, Marketing Engineer, feels that "over the long haul, the nation will be turning to coal." Thus, the market for Fluidyne's fluidized-bed systems will develop, though he admits that because the economics and reliability of fluidized-bed air heaters have not been demonstrated, "right now, industry is not beating a path to our door."

Sirany sees the lack of an established transportation and coal-handling network to deliver coal to small users as a major barrier to the successful commercialization of fluidized-bed products. He feels that, until this network is developed, it will be difficult for fluidized-bed systems to make any **real** dent in the market for small-scale industrial systems.

Fluidyne is offering its fluidized-bed air heaters as commercially available products. It considers itself the only company in the world which can offer a demonstrated fluidized-bed system to heat air. Thus far, Fluidyne is still awaiting its first commercial offer.

Address

Fluidyne Engineering Corporation
5900 Olson Memorial Highway, Minneapolis, Minnesota 55422

Financial Data

Fluidyne Engineering Corporation is privately owned; no financial data is available.

FOSTER WHEELER CORPORATION

Summary

With sales of over $1 billion, Foster Wheeler is a multi-national corporation engaged in the design, engineering, fabrication and construction of steam generators and process plants. Foster Wheeler's interest in fluidized-bed combustion began in the late 1960s, and the company has played a major role in United States efforts to prove the technology economic. Foster Wheeler is commercially offering industrial fluidized-bed units capable of producing up to 600,000 pounds of steam an hour.

Program and Technology

In 1969, experiments with the burning of lignite and other low-grade fuels led Foster Wheeler to fluidized-bed combustion. The sulfur-capturing possibilities of the technology soon became obvious. Foster Wheeler joined with the consulting engineering firm of Pope, Evans and Robbins Incorporated (PER) (see PER profile) to develop fluidized-bed combustion in response to the sulfur-dioxide emission standard promulgated by the U.S. Environmental Protection Agency (EPA). The partners designed an 800-MWe "multi-cell" fluidized-bed combustion unit (see PER profile) consisting of four identical 200 MWe segments stacked vertically.

In October 1972, PER signed a $30 million contract with the U.S. Office of Coal Research (OCR, later part of the U.S. Energy Research and Development Administration [ERDA], now part of the U.S. Department of Energy [DOE]) to design, construct and operate a 30 MWe atmospheric fluidized-bed boiler to operate under practical utility conditions at the Monongahela Power Company's plant in Rivesville, West Virginia. The plant was designed to provide the experience and information necessary for a scale-up to 800 MWe. PER subcontracted the design, fabrication and construction of the boiler to Foster Wheeler.

Foster Wheeler operates two fluidized-bed testing facilities in Livingston, New Jersey: a 1.75 square-foot bed area pilot unit for component testing and process evaluation, and a 36 square-foot unit. These facilities, along with PER's

Alexandria test boiler, provided the design specifications for the Rivesville unit.

Foster Wheeler has developed an "overbed" feed system which feeds coal into the bed from above. It feels this system will deal successfully with the feed problems at the Rivesville plant (see PER profile). DOE has approved installation of part of this new feed system but wants the plant to log more operating hours with the present system before installing a new one. Foster Wheeler feels that uncertainty over the reliability of coal-feed systems is discouraging many potential customers.

Foster Wheeler is now offering commercial fluidized-bed industrial boilers capable of producing up to 600,000 pounds of steam per hour. These units will be covered by a commercial warranty. It has submitted numerous proposals to potential customers for fluidized-bed boilers, but has thus far signed no contracts. Fluidized Combustion Corporation (FCC), a subsidiary of Foster Wheeler and PER, has been awarded several contracts. For the Tennessee Valley Authority (TVA), FCC is preparing a preliminary design study for a 150 MWe atmospheric fluidized-bed boiler to burn the lowest grades of coal. (TVA has also contracted for parallel studies with Babcock & Wilcox [U.S.] and Combustion Engineering.) FCC is working jointly with the Battelle Memorial Institute to build a 25,000 pound-per-hour "multi-solid fluidized bed combustion" boiler for DOE's industrial demonstration fluidized-bed program (see Battelle profile). FCC is the subcontractor to Georgetown University for another industrial demonstration unit, a 100,000 pound-per-hour boiler.

FCC has acquired 51 percent and Foster Wheeler an additional 25 percent of Copeland Systems Incorporated of Oak Brook, Illinois, an engineering concern specializing in fluidized-bed combustion technology. FCC has signed licensing agreements for European rights to its atmospheric fluidized-bed systems with Foster Wheeler Power Products, Ltd. (U.K.), a wholly owned subsidiary of Foster Wheeler Corporation, and with AB Generator of Sweden.

Foster Wheeler has contracted with the architect engineering firm Stone & Webster, Inc., for the design of a 570 MWe atmospheric fluidized-bed steam-generation system for DOE. It has also submitted two other proposals: to build DOE's Atmospheric Fluidized-Bed Component Test and Integration Unit (CTIU) at the Morgantown Energy Research Center in Morgantown, West Virginia, and to design and construct a steam-generating and process-heating fluidized-bed facility burning anthracite tailings.

Goals and Opinions

Foster Wheeler's ultimate goal is to prove its atmospheric fluidized-bed combustion systems economic and market them worldwide. Its immediate goal is to demonstrate an overbed feed system successfully at the Rivesville plant.

According to Robert Gamble, Manager of Foster Wheeler's Development Engineering Department, the foreign market for fluidized-bed systems may open up before the domestic market. He also feels that, because a fluidized-bed system has fewer functioning parts than a conventional boiler with a scrubber, the fluidized-bed system will prove more reliable and economical. Gamble says that capital costs for fluidized-bed systems will be lower than those of a conventional pulverized-coal system with a scrubber and the operating costs will be a "standoff." "Once the calcium/sulfur ratio for the fluidized bed is optimized, the economics will swing even more in its favor."

Gamble sees two courses which the government could follow in fluidized-bed combustion. It could fund research and development efforts and demonstration plants until the technology is fully commercialized at both the industrial and utility level. Alternatively, it could let the technology take its own course, slowly developing from a few industrial units until, at some later date, utilities gained enough confidence in the technology to install it. Gamble feels that the latter path would take at least twenty years and that the former would prove the technology economic sooner.

Address

Foster Wheeler Corporation
110 South Orange Avenue, Livingston, NJ 07039

Financial Data

(figures in thousands)	1977	1976
Sales	$1,189,007	$1,061,723
Net income	27,063	20,548

NATIONAL COAL BOARD

Summary
The National Coal Board (NCB) owns and operates the coal industry of the United Kingdom. The NCB was formed by an Act of Parliament in 1947 as a government-owned "nationalized industry." At tnat time coal was the predominant fuel in the United Kingdom. A decade later, imported oil had begun to displace coal. The discovery of offshore natural gas and oil further reduced coal's share of the United Kingdom's fuel market. By the late 1960s, against the NCB's advice, United Kingdom government policy was to phase out the coal industry as rapidly as possible. However, after the rapid rise in oil prices in 1973, coal underwent a renaissance in the United Kingdom. The government, the NCB, and the National Union of Mineworkers agreed on a "Plan for Coal" calling for massive investment in new coal faces and new mines. The plan provided renewed support for research and development, focusing particularly on the NCB's work on fluidized-bed combustion at its two major laboratories.

Program and Technology
The NCB carries on an extensive program of research and development in coal technology, although it does not manufacture equipment. It operates two coal-utilization laboratories, one at Stoke Orchard, Gloucestershire, the other at Leatherhead, Surrey. Together they are known as the Coal Research Establishment (CRE).

Work on fluidized-bed combustion is divided between the two laboratories. Fluidized-bed combustion work at Stoke Orchard, which takes a major part of the Ł 5.5 million annual research and development budget of the CRE, focuses on atmospheric systems. Stoke Orchard employs about 535 people, including 140 scientists and engineers. Work at Leatherhead focuses on pressurized systems, and is almost entirely self-supporting via contract research. Leatherhead has a staff of 70.

CRE began research on fluidized-bed combustion in 1966. At Stoke Orchard, a 1 MW (thermal) test rig and a number of smaller rigs have been operating for some years. They have provided design data on various types of boilers, incinerators, and agricultural crop dryers. All these applications have since come into commercial use in the United Kingdom. A large municipal waste-disposal atmospheric fluidized-bed unit is now operating in Caernarvon, Wales. Coal-fired fluidized-bed crop dryers are now being sold commercially. Several companies now offer fluidized-bed boilers and conversion kits based on CRE specifications. A 10,000 pound-per-hour boiler at a tomato nursery in Herefordshire, converted to coal-fired fluidized-bed firing, has been in operation since early 1977. A fluidized-bed modification of a similar boiler at a leatherworks in Lancashire is nearing completion.

On a much larger scale, CRE designed the modification of an 80,000 pound-per-hour boiler at the British Steel Corporation River Don plant in Sheffield, Yorkshire. A coal-fired fluidized-bed combustor is being installed in place of an oil-fired burner. Work began in the summer of 1977, and the remodeled boiler is expected to be in service by the spring of 1978. The steam from this atmospheric fluidized-bed boiler will drive hammers, generate electricity, and provide steam for space heating and process purposes. This installation will be the world's first industrial-scale coal-fired fluidized-bed cogeneration unit.

CRE data also provided the basis for the 13 MW (thermal) atmospheric fluidized-bed modification carried out by Babcock and Wilcox Limited (B & W [U.K.]) at its site in Renfrew, Scotland (see Babcock and Wilcox [U.K.] profile). The NCB and B & W (U.K.) have a close working relationship, described below.

The Leatherhead laboratory operates an 8 MW (thermal) pressurized fluidized-bed test rig, the largest of this type in the world. It has been in service since 1969. The Leatherhead unit has provided basic design information for a number of United States fluidized-bed projects (see American Electric Power Company, Stal-Laval Turbin AB, Babcock & Wilcox Limited and Curtiss-Wright Corporation profiles), and for the International Energy Agency Grimethorpe project (see NCB [IEA Services] profile).

Goals
and
Opinions

Dr. Joseph Gibson, NCB Board Member for Science and the person responsible for the entire program at the two laboratories, puts the NCB's primary objective succinctly: "We want to sell coal." The NCB wants to encourage the use of advanced coal technologies in order to enlarge the market for its basic product. The NCB does not, therefore, want to pursue particular financial advantage from development of coal-utilization technology itself. This attitude strongly influences the NCB approach to patents and licenses for hardware and design data developed by CRE. In 1973, the NCB entered into partnership with British Petroleum (BP) and with the National Research and Development Corporation, a government-financed research and development organization, to form Combustion Systems Limited (CSL). CSL's main objective was to commercialize fluidized-bed combustion (see CSL profile). The NCB actively encourages development by providing financial support, for example, for construction of the River Don plant and the boiler modification at the tomato nursery. The nursery is, however, running the boiler and paying its operating costs. The NCB will also run the River Don unit at first, although the British Steel Corporation will pay its operating costs. British boiler manufacturers, including Mitchell Engineering, Worsley, and Babcock & Wilcox Limited (U.K.), have arrangements with the NCB through CSL for use of fluidized-bed combustion data. Mitchell has been building the River Don unit in their shop, to NCB specifications, before installing it on site. NCB will "carry the can" in the event of any problem at River Don, though the agreement is not in the form of a conventional warranty.

The NCB acknowledges that, for the next decade or more, their primary objective of selling more coal will be far from easy to accomplish. The NCB's major customer is Britain's Central Electricity Generating Board (CEGB), which burns about 70 million tons of the NCB's 120 million-ton annual output. The CEGB, however, since the late 1950s and until recently, has grown less and less enamored of coal. Over-ordering of new CEGB plants -- mainly oil-fired and nuclear -- in the 1960s has left the CEGB with an embarrassing excess of generating capacity. Accordingly, no new base-load generating plant has been ordered since 1972, and the CEGB is now opposed to ordering any new plant of any kind. The British Government has nevertheless just required the CEGB to order a new coal-fired pulverized-fuel baseload station at Drax, in Yorkshire, to provide work for the struggling power-plant construction industry. Also being discussed is a similar premature order for a nuclear plant. The government will compensate the CEGB for the additional cost of a premature order in each case.

The NCB has an interest in the Drax order. But many NCB officials feel that a government subsidy to construct at least one fluidized-bed demonstration plant for electricity generation would have been timely, possibly in conjunction with combined cycles or district heating. As it is, the excess of existing generating capacity makes it unlikely that any full-scale fluidized-bed plant will be ordered in the near future. The NCB, therefore, plans to devote its immediate attention to developing industrial- and commercial-size fluidized-bed systems. Its goal for such development is to win back some of the old markets for coal. These markets, however, are at present supplied by offshore natural gas, for the moment both plentiful and cheap, and oil. The people at the NCB acknowledge that, for the next decade, because of North Sea oil and gas, their primary objective of selling more coal will be far from easy to accomplish. But their morale is higher than it has been for twenty years. They are convinced that fluidized-bed combustion will be adopted as the main method of burning coal and other fuels. They can fairly claim that their pioneering work has done a great deal to make this possible.

| Address | National Coal Board |
| | Hobart House, Grovenor Place, London SW1 XAE, England |

Financial	(figures in thousands)	1976
Data	Sales	£48,400
	Net income	500

NCB (IEA SERVICES) LTD.

Summary

To respond to the worldwide 1973 oil-price increases, the countries of the Organization for Economic Cooperation and Development (OECD), the major economic association of Western industrial countries, formed the OECD International Energy Agency (IEA). One of the IEA's objectives was international cooperation in development of advanced coal technology. The United Kingdom was made lead country for the IEA's coal-technology program because of the United Kingdom's long involvement and experience in coal research. The British government asked the National Coal Board (NCB) to assume responsibility for the IEA programs. Leslie Grainger, then NCB Board Member for Science, chaired the IEA Coal Working Group.

Five projects were undertaken. Four were designed to draw together information and make it available. The fifth was a joint demonstration project to build a large-scale prototype pressurized fluidized-bed test rig at Grimethorpe, Yorkshire. Participating countries were the United Kingdom, the United States and West Germany. To manage and operate this and the other projects, the United Kingdom set up NCB (IEA Services) Ltd., a wholly owned subsidiary of the NCB, with Grainger as chairman. On November 20, 1975, international agreements were signed between the countries involved, with each providing one-third of the total Ł17 million funding. The Grimethorpe project is the largest pressurized fluidized-bed project under construction in the world today.

Program
and
Technology

The NCB had been prepared to build a fluidized-bed prototype at the Grimethorpe site since 1968, though it had not been able to get government funds to do so. The NCB had originally planned to build an atmospheric fluidized-bed unit. However, the plan as put forward to the IEA was for a flexible pressurized rig, heavily instrumented and therefore capable of wide-ranging modification for testing of components and parameters.

The IEA's Coal Working Group accepted the British plan, and the rig is now under construction. It is expected to be commissioned by 1979. A four-year experimental program on the various aspects of the rig will follow. When it comes into service, it will be the largest pressurized fluidized-bed rig in the world, with a nominal output of 80 MW (thermal). The rig will not be coupled to a turbine. IEA Services says that flexibility will thus be greater. However, other fluidized-bed experts consider that enough is already known about the basic design data to proceed directly to a unit with a turbine. (See profiles on the Curtiss-Wright Corporation, Stal-Laval Turbin AB and Babcock & Wilcox Limited [U.K.].) There is room for a gas turbine on the Grimethorpe site, if in due course it is decided to add one.

Design of the rig is now virtually complete. All major construction contracts were placed by the end of October, 1977. Vereinigte Kesselwerke AG, Dusseldorf, will build the pressure vessel and Petrocarb, Inc., New York, the coal-feed equipment. The British firms of Head Wrightson Process Engineering Ltd., Miller Construction Ltd., Babcock Controls Ltd. and Compair Industrial will build the remaining components of the plant and do the site engineering. Fabrication and site work have been underway since early 1977. Capital costs for the Grimethorpe plant are expected to total about £9 million, and the other costs, including those of the experimental program, will bring the total outlay to £17 million. The project will be financed by the governments of the three participating countries.

Goals
and
Opinions

The primary goal of IEA Services for the pressurized fluidized-bed Grimethorpe rig is to complete construction, commission the unit, and launch the planned experimental program. United States participation is administered by the U.S. Department of Energy (DOE). United States corporations will obtain Grimethorpe data by sending staff to work at Grimethorpe. Staff from corporations will be selected by a committee of the American Society of Mechanical Engineers (ASME). West German participation is administered by the Juelich Nuclear Research Center.

There is already a possibility that other OECD countries, originally reluctant to take part, will in due course buy themselves into the project. Patent questions will be settled by the three governments involved. British rights will be available through Combustion Systems Limited, the

NCB fluidized-bed partnership, probably in the form of licensing or consultancy agreements (see CSL profile).

The NCB itself is now somewhat ambivalent about its involvement in the IEA Services work, partly because of its large financial commitment. Some NCB officials also feel that Britain will be giving too much and getting too little in return on the technical side. There are other political complications in the international venture. Orders for hardware must be distributed among the participating countries, on a basis of perceived fairness, which may run counter to the usual technical and economic criteria. Management and administration decisions must be made with an eye to political implications. Elsewhere in the IEA, there is evidence that such international hardware projects are very difficult to implement. Grainger is convinced that only the NCB's readiness with a detailed plan in 1975 made the Grimethorpe project possible. "We'd never get away with Grimethorpe again."

Be that as it may, Grimethorpe is the only large-scale pressurized fluidized-bed combustion project which has come so far. Its progress will be closely watched.

Address   National Coal Board (IEA Services) Ltd.
Hobart House, Grovenor Place, London SW1 X7AE, England

Financial   No financial data available.
Data

OAK RIDGE NATIONAL LABORATORY

Summary   Oak Ridge National Laboratory (ORNL), located in Oak Ridge, Tennessee, is operated for the U.S. Department of Energy (DOE) by the Nuclear Division of the Union Carbide Corporation. ORNL was established during World War II for nuclear research and development. Recently it has broadened the scope of its activities, and is now carrying out a contract with DOE on fluidized-bed combustion.

Program
and
Technology

Under a contract with the Department of Housing and Urban Development (HUD) and the Office of Coal Research (OCR, later part of the Energy Research and Development Administration [ERDA], now part of DOE), ORNL initiated development of a "Modular Integrated Utility System" (MIUS). This is a small-scale "total energy" installation -- a system which supplies both electricity and heat for water and space heating -- for use in urban housing developments.

In 1972, HUD asked ORNL to investigate a coal-fired MIUS. The OCR provided technical direction. After looking at a wide variety of systems, ORNL, HUD and ERDA decided that an atmospheric fluidized-bed combustor coupled to a gas turbine, would be a promising way to burn coal cleanly in an urban context. The ORNL developed such a system, which is unique in coupling these two elements. The ORNL design differs from other fluidized-bed/gas-turbine systems in that the fluidized-bed system is atmospheric, not pressurized. The hot gases from the burning coal are discharged through clean-up systems; they do not enter the gas turbine itself. Instead, the hot bed heats air to 1500°F in pressure tubes. This clean hot high-pressure air is then fed through the gas turbine at a lower temperature (1050°F) and pressure and goes to a set of heat exchangers, which cool the air and extract waste heat from it. (The waste heat is recovered at a high temperature and thus can be used for industrial process applications.) The air is sent from the heat exchangers to a compressor and then fed back into the tubes in the fluidized bed to be reheated to working temperature for another cycle through the turbine. The closed-circuit air flow through the tubes in the fluidized bed and the turbine eliminates what would otherwise be the serious problem of removing particulates and other contaminants.

Design work on the MIUS continued until September 1976, after which ERDA redirected the project toward developing a fluidized-bed combustion design for industrial applications. ORNL has now produced a concept for an atmospheric fluidized-bed Coal Combustor for Cogeneration Technology Test Unit (CCC). The CCC would be able to produce 300 to 500 KW of electricity, plus some 2.5 million Btu-per-hour of useful heat. Boiler-design firms are being asked to submit designs for the combustor, and ORNL will choose a final design from among their proposals.

Arthur Fraas, former director of the project and now a consultant to the laboratory, points out that ORNL has tested coal-feeding and erosion/corrosion problems extensively. This testing has convinced him that by feeding dry coal, by carefully metering the coal's arrival in and dispersion throughout the bed, and by accurately monitoring and maintaining the correct amount of excess air in the bed, fluidized-bed systems can be operated without the major problems of erosion/corrosion of in-bed tubes and fuel feed.

Goals and Opinions

ORNL's basic goal for the CCC program is to prove the suitability of a fluidized-bed closed-cycle gas-turbine unit for large industrial applications. ORNL has estimated the performance, reliability and erosion/corrosion characteristics of the fluidized-bed combustor. It has also tested coal and limestone handling and feed systems and has produced specifications for a boiler. Robert Holcomb, director of the project, says the project is ready to proceed to the final design and construction phases.

Fraas feels that gas turbines operated on the combustion products from pressurized fluidized-bed units, at the temperatures the units require to operate (1600°F), will have severe problems with erosion and deposits on the turbine blades. He says "This has been the case in many previous efforts to operate coal-fired gas turbines; hence this approach will probably not prove practicable."

Address

Oak Ridge National Laboratory
Oak Ridge, Tennessee 37830

Financial	(figures in thousands)	1977	1976
Data	Research volume	$229,000	$194,000

POPE, EVANS AND ROBBINS INCORPORATED

Summary

Pope, Evans and Robbins Incorporated (PER), is a consulting engineering firm based in New York whose interest in coal dates back to the 1940s. PER began to investigate fluidized-bed combustion in 1962. Such work at PER now involves some eighty scientists and engineers, making it one of the most active organizations in the field. "We've bet our company on it, says Michael Pope, PER Chairman and Chief Executive Officer. A major part of the company's efforts center on a 30 MWe demonstration plant in Rivesville, West Virginia, currently the world's largest operating fluidized-bed combustion facility, and a 100,000 pound-per-hour industrial boiler demonstration plant at Georgetown University, Washington, D.C.

Program and Technology

In 1965, with funds from the U.S. Office of Coal Research (OCR, formerly with the U.S. Energy Research and Development Administration [ERDA], now the U.S. Department of Energy [DOE]), PER set up a fluidized-bed laboratory at Alexandria, Virginia. The first testing started up there in 1965. The third and largest boiler built there is the present 0.5 MW (thermal) fluidized-bed unit which has been used to test a wide range of fuels and operating modes.

In October 1972, PER signed a $30 million contract with OCR to design, construct and operate a 30 MWe prototype atmospheric coal-fired fluidized-bed boiler at Rivesville, West Virginia. The object at the time was to develop a coal-fired boiler for power generation which would be cheap enough to compete with existing oil-fired systems. The boiler was to be installed in the existing Rivesville power station of the Monongahela Power Company, part of the Allegheny Power System. The boiler was built to PER specifications by the Foster Wheeler Corporation, and general construction was done by Champion Construction and Engineering, Inc.

The design for the Rivesville plant called for a "multi-cell" fluidized-bed unit, consisting of four separate fluidized beds side by side. In effect, the first fluidized bed boils the water, the second heats the resulting steam, and the third superheats the steam. These three beds operate at 1550°F for optimal sulfur control. The combustion gases from these three cells carry away ten to fifteen percent of the coal as unburned particles. This coal is captured in cyclone collectors and fed into the fourth fluidized bed, which operates at a higher temperature (2000°F) and a slower fluidizing velocity. This cell is known as the "carbon burn-up" cell. The multi-cell design is a controversial feature of the Rivesville plant. Other fluidized-bed experts have questioned its necessity or advisability. PER feels that the multi-cell design offers advantages in start-up, control of final steam temperature, and overall combustion efficiency.

As might be expected for a prototype, the Rivesville plant encountered a variety of problems. Space limitations on the Rivesville site made it necessary to have the four cells side by side instead of one on top of the other, as originally intended. Some components proved unsatisfactory. Replacing them under warranty contributed to a lengthy delay in commissioning the plant. At one stage, Rivesville was expected to be operating by July 1976. However, the first successful firing of one cell with coal did not take place until December 7, 1976, after protracted experiments with start-up procedures. The first operation of all four cells took place on August 20, 1977, and the first commercial operation occurred on September 16, 1977. With three cells in operation, the plant was used to generate 13 MWe. According to PER, "The Rivesville unit thus became the world's first fluidized-bed boiler to generate electricity."

The plant is still, however, having difficulties, especially with coal handling and coal feed. A combination of circumstances led to a persistent problem of clogging of coal in feed lines, blocking the fuel supply. PER considers that the fluidized-bed system itself has proved satisfactory. But the fuel-feed problems persist. Further major modifications of the fuel system are now underway.

Goals
and
Opinions

PER's long-term goal is to design and sell fluidized-bed boilers to commercial users. However, its more immediate aim is to operate the Rivesville plant as a working power plant, resolving problems of materials, maintenance and control.

In addition to Rivesville, PER has done feasibility studies for about a dozen other projects. It has licensed Mitsubishi and IHI of Japan to build a 300,000 pound-per-hour atmospheric fluidized-bed boiler to burn "coke breeze," a low-quality fuel usually regarded as waste. PER is an equal partner with Foster Wheeler in the Fluidized Combustion Company (FCC), which has been awarded contracts by the Tennessee Valley Authority (TVA), Battelle Memorial Institute, Georgetown University and major corporations. TVA has contracted for preliminary design studies of a 150 MWe atmospheric fluidized-bed boiler to burn the lowest grades of coal. (TVA has also contracted for parallel studies with Babcock & Wilcox [U.S.] and Combustion Engineering.) FCC is building a 100,000 pound-per-hour boiler at Georgetown University under a contract with DOE. The boiler is expected on stream by the end of 1978.

FCC has licensed Foster Wheeler Power Products, Ltd., of the United Kingdom and AB Generator, of Sweden, to use its atmospheric fluidized-bed systems in Europe. FCC has also acquired 51 percent of Copeland Systems, Incorporated, of Oak Brook, Illinois, which has built some 100 fluidized-bed incineration and waste-heat installations. However, PER has "no desire to expand," according to Michael Pope. The company is negotiating an agreement with Stone & Webster Engineering which provides for the joint marketing of consulting engineering services for industrial fluidized-bed boiler plants. PER and Stone & Webster are already collaborating on conceptual designs for a 570 MWe atmospheric fluidized-bed power plant.

PER intends to concentrate on atmospheric, rather than pressurized, fluidized-bed systems. In 1962, Michael Pope did a detailed paper analysis of the two types of systems which showed that there was, at most, a 10 percent improvement in cycle efficiency using pressurized fluidized-bed technology, and that the costs would not be very different. However, fuel-feed problems would be much greater in a pressurized system. Feeding coal into a boiler is always difficult, and to do so under pressure, Pope says, "I wouldn't wish on my worst enemy."

Pope feels that customer uncertainty about warranties will continue to be a stumbling block in the development of fluidized-bed technology. Recent announcements of both Foster Wheeler and Babcock & Wilcox Limited (U.K.) that commercial warranties are now being offered for industrial-size boilers should do much to accelerate commercialization.

Pope believes that government funding of up to three central electricity-generating stations using fluidized beds would also accelerate the commercialization of the technology.

Pope is satisfied that the disposal of limestone sorbent (which removes sulfur) will not cause problems. Unlike scrubber sludge, which is a viscous and biologically pernicious liquid, fluidized-bed waste material -- spent limestone sorbent and coal ash -- is a dry powder, biologically innocuous if not actually beneficial. Pope hands out packets of peanuts grown on soil conditioned with sulfated limestone from a fluidized-bed unit. He believes that the solid fluidized-bed residue will be not a waste but a by-product.

PER's original objective was a coal-burning boiler that did not pollute and that was cheaper to build and operate. Pope says, "I was looking for motherhood." He is convinced that, with atmospheric fluidized-bed combustion, he has found it.

Address	Pope, Evans and Robbins Incorporated 1133 Avenue of the Americas, New York, New York 10036		

Financial Data	(figures in thousands)	1977	1976
	Sales	$9,353	$10,778
	Net income	140	119

STAL-LAVAL TURBIN AB

Summary   Stal-Laval Turbin AB is a Swedish heavy-engineering firm
with subsidiaries in six other countries. The company itself
is a wholly owned subsidiary of Allmänna Svenska Elektriska
AB (ASEA). Stal-Laval's 1977 sales were $183.8 million;
its parent company's sales were $808 million.

Stal-Laval's major product is turbine plants, including both
steam and gas turbines for utility, industrial and marine
use. In 1969, Stal-Laval Limited, the British subsidiary,
initiated interest in developments incorporating fluidized-
bed combustors.

Program
and
Technology

In late 1969, Henrik Harboe, now Managing Director of Stal-
Laval Limited, happened on an article by Professor Douglas
Elliott, which described a pressurized fluidized-bed unit
that could be coupled to a gas turbine. At Harboe's instigation, Stal-Laval began to investigate possible involvement in
the technology. In the spring of 1973, Harboe published a
paper entitled "The Importance of Coal," which discussed
fluidized-bed combustors as a component of design proposals for total-energy systems. In response, in the summer
of 1973, Woodall-Duckham Limited (see Babcock & Wilcox
[U.K.] profile) contacted Stal-Laval. By the end of that
year, Harboe had brought about agreement between Stal-
Laval and Woodall-Duckham to cooperate on development
of systems coupling Woodall-Duckham fluidized-bed units
to Stal-Laval turbines.

The next step was for the partners to find a customer for
their concept. In the United Kingdom, neither the National
Coal Board, the Department of Industry nor the Central
Electricity Generating Board were interested. In the United
States, in 1975, the U.S. Energy Research and Development
Administration (ERDA, now U.S. Department of Energy
[DOE]) issued "Requests for Proposals" for development of
advanced coal technologies. Though Stal-Laval and its long-
time United States associate Pratt & Whitney put in a bid,
they lost out to the Curtiss-Wright Corporation (see Curtiss-
Wright profile). However, in October and November 1976,
the American Electric Power Service Corporation (see Amer-

ican Electric Power Company [AEP] profile) expressed an interest in the fluidized-bed gas-turbine system which Stal-Laval and Woodall-Duckham had been investigating. Discussions led to the first phase of a design study, which as of early 1978 was nearing its conclusion. AEP, Stal-Laval and Woodall-Duckham have carried out two programs of test operation on the pressurized fluidized-bed rig at the National Coal Board's Leatherhead laboratory. The rig is burning AEP coal in order to study turbine-blade effects and emission control. If the results prove satisfactory, the second phase of the project will be a detailed engineering design for a 170 MWe combined cycle plant. Stal-Laval hopes for an actual order early in 1979.

Stal-Laval's interest in marine turbines has prompted another line of development. To get hotter, higher-pressure steam from conventional oil-fired marine boilers, Stal-Laval has proposed adding fluidized-bed super-heaters and reheaters burning heavy residual fuel oil. At a recent London symposium on advanced steam-turbine technology at the Institute of Marine Engineers, the Stal-Laval plans for "Very Advanced Propulsion Systems" attracted considerable attention. Stal-Laval hopes that such advances may also lead to better land-based systems, and possibly in due course to complete fluidized-bed boiler systems for marine propulsion burning either heavy oil or coal.

Goals
and
Opinions

As a turbine manufacturer, Stal-Laval is interested in developing new and more efficient applications of such technologies, over a range of sizes, for both gas and steam turbines. The versatility of fluidized-bed combustion technology lends itself to such advanced turbine applications as combined-cycle systems and co-generation of electricity and heat. Harboe considers, however, that utilities are unlikely to order atmospheric fluidized-bed systems. In his view, such systems are proving physically much larger than anticipated and offer no advantages in cost or efficiency. For an atmospheric system, to reduce size relative to output involves other technical measures which can create considerable problems. For example, using a higher fluidizing velocity may require adding a carbon burn-up cell (see Pope, Evans and Robbins Incorporated profile). Harboe says this means "doing what you set out to avoid": moving to higher temperatures and coping with other troublesome engineering parameters.

Harboe is certain that pressurized fluidized-bed systems will be cheaper, and that, far from adding complications, feeding coal in such a system will be easier than in a large atmospheric system. Harboe expects pressurized fluidized-bed combustion to have a much wider field of application for both industry and utilities. "You want to convert people who are big users of oil and gas to using coal. How do you convince them? You show them that they can backfit a pressurized fluidized-bed unit, and simultaneously double their generating capacity by moving to combined cycles." He sees such backfitting especially for industrial applications, using a topping cycle plus conversion to coal-fired fluidized-bed combustion. Harboe is a vigorous enthusiast for total energy systems. He is convinced that they must be coal-fired, and that fluidized-bed combustion is the key to making them clean, efficient and economic.

| Address | Stal-Laval Turbin AB |
| | S 612 20 Finspong Sweden |

| Financial | (figures in thousands) | 1977 | 1976 |
| Data | Sales | $203,820 | $143,840 |

STONE-PLATT FLUIDFIRE LIMITED

Summary   Fluidfire Development Limited was established in 1972 by Douglas Elliott and his fellow engineer Michael Virr. In the subsequent six years, Fluidfire has accumulated a record second to none in imaginative development of fluidized-bed technology, and in commercial marketing of its products. Its original aim was commercial development of Elliott's fluidized-bed innovations, particularly those involving shallow beds. As this research company's work progressed, three additional companies were set up to market the different product lines developed. Fluidfire Furnaces Limited makes fluidized-bed furnaces for heat treatment. Fluidfire Waste Heat Systems Limited markets a novel type of fluidized-bed heat exchanger, one of which has been installed on a Norwegian tanker. Tolltreck Fluidfire Limited manufactures industrial fluidized-bed incinerator-boilers.

Although Elliott died in mid-1976, Fluidfire has continued to expand. The majority interest in the company now belongs to Stone-Platt Industries, a British engineering group. Virr continues as managing director of what is now called Stone-Platt Fluidfire Limited, which includes the three subsidiaries.

Program
and
Technology   Fluidfire Development began by marketing fluidized beds for metallurgical heat treatment, a traditional application. The company then embarked on a new approach to fluidized-bed design, in which the bed was only a few inches deep even when operating, compared to a depth of a foot or more in conventional beds. Fluidfire introduced a range of heat exchangers and waste-heat boilers utilizing the shallow-bed concept.

The heart of Fluidfire's design is a shallow bed with one or more boiler tubes running horizontally just above its surface. The tube is positioned so that when the bed is "slumped" -- when the layer of particles is motionless on the floor of the chamber -- the tube is above the surface of the bed. When the bed is fluidized -- that is, when the air flow has lifted the particles into motion -- the surface of the "bed" rises until the tube is submerged. If desired, the velocity of the fluidizing air may be adjustable to allow the bed to rise only partway up the boiler tube. In this way the heat-transfer characteristics of the system can be varied.

Most applications of fluidized-bed technology for energy supply involve burning fuel in the bed. However, the simplest application of Fluidfire's shallow-bed concept does not involve combustion in the bed. Instead, the bed serves merely as a heat-transfer medium. A shallow-bed heat exchanger or waste-heat boiler uses hot exhaust gases -- perhaps from a diesel engine -- to fluidize a shallow bed. (The shallow bed does not require much pressure to fluidize, avoiding back-pressure on the engine exhaust.) The hot gases transfer their heat with high efficiency to the particles of the bed. The bed in turn transfers the heat with similar high efficiency to boiler tubes. In a conventional heat exchanger, the hot gases must transfer their heat directly to boiler tubes. The use of the fluidized-bed as an intermediary greatly facilitates the overall heat transfer. The system can thus be made much more compact for a given output. Fluidfire has built and delivered a number of such units, including a design for a package waste-heat exchanger delivering 250,000 to 5,000,000 Btu-per-hour. This is capable of handling flue gases at temperatures higher than $1000^{o}C$, and has fully automated control systems.

The most recent innovation in this line is an installation now operating on the 40,000-ton Norwegian tanker _Fjordshell_. The _Fjordshell_ boiler uses three shallow beds, one above the other, fluidized by the exhaust gas from the ship's 12,000-horsepower diesel engine. Each bed is nine by twelve feet. The whole unit is some eighteen feet high. The boiler underwent demanding sea trials at the end of 1977 off Norway's northernmost coast above the Arctic Circle and performed to the designers' expectations. The turbulent sea did not affect the behavior of the operating beds, and the boiler output of 7,000 to 8,000 Btu-per-hour of saturated steam saved the ship operators some $8,000 per month in oil costs. The dirty exhaust gases from heavy fuel oil posed no problem after an automatic mechanical scrubber was installed to clean the interior of the heat exchanger. Fluidfire is now offering similar units producing up to 5 MWe output for co-generation and marine use.

Fluidfire was similarly innovative in developing fluidized beds for waste incineration. A problem encountered in early attempts to use fluidized-bed designs for waste incinerators was how to keep the waste in the bed long enough to burn it completely. Some kinds of combustible material tended to be lifted out of the bed before they had been burnt. Fluidfire designed an ingeniously simple way to increase the "residence time" of combustible material fed into a fluidized-bed burner. Instead of using a chamber with a horizontal plane floor, the Fluidfire design uses a chamber in the form of a long trough, with a cross section like a wide curving V. The fluidizing air enters through the entire porous fire-brick wall area of the trough. As a result the bed material rises along the walls, flows inward from the upper edges to the center line of the surface of the trough and descends from this center line to the bottom center line of the trough. It then separates and flows up again along each wall. Fuel or waste is delivered by a screw feeder into the center line of the bed and is carried down by the currents of the descending bed material. This initial downward motion offsets any tendency for light fractions of fuel or waste to be blown out of the bed. The feed design also eliminates any problem of clogging of feed lines, making it possible to burn wet coal or other awkward materials.

The residence time thus achievable makes it possible to burn low-grade fuels like high-ash coal completely. The Fluidfire design does not include boiler tubes in the main bed. Instead, there are a pair of shallow-bed heat exchangers at the outer upper edges of the trough. When the main bed has reached operating temperature, the shallow beds are fluidized. The main bed then exchanges hot bed material with the shallow beds, and heat is transferred to the boiler tubes in the shallow beds.

Fluidfire has just completed construction of a 1 million Btu-per-hour coal-fired boiler of this design for the coal workshop of the Virginia Polytechnic Institute and State University, which is directed by Professor Arthur Squires. Squires will carry out a program of research on this system to investigate its preformance and emission-control characteristics burning very high-sulfur coal.

Two other Elliott inventions are still in an early development stage and open up entirely new areas for fluidized-bed applications. One is designed for use in homes. It is based on a very shallow bed of very small area, small enough to fit into the ordinary domestic fireplace. A prototype of such a fluidized-bed fireplace was built at Elliott's laboratory at Aston University. The fluidized-bed home unit burns coal very slowly, at less than one kilogram per hour, but produces a high heat output. Measurements indicate that the bed converts some 50 percent of the coal energy to radiant heat, compared to only 15 percent in an ordinary fireplace. A gas-fired version has already been tested in a model home in the Netherlands.

Elliott's other innovation is at an even earlier stage. It is a different approach to the velocity limit on heat rate (see Battelle profile). The Fluidfire design uses a spinning cylinder whose wall is lined with a fluidized-bed, held in place by the centrifugal force produced by the rotation of the cylinder. (The effect is similar to that of the fairground Wall of Death ride.) The faster the cylinder rotates, the more the bed particles resist being blown radially inward by fluidizing air from the outer wall of the cylinder. The effect is similar to that produced by the density of the hematite in the Battelle multi-solid fluidized bed. But in the Fluidfire rotating bed the forces restrain the coal and ash particles themselves, making possible a much higher rate of heat release per unit area of bed. A bench-scale model of the Fluidfire rotating bed has produced heat output comparable to that from the core of an operating fast breeder reactor, some 500 kilowatts per litre.

Goals and Opinions

Fluidfire is still a small firm, especially when compared to other firms profiled in this report. It employs 26 people. Until recently, lack of capital severely constrained expansion. Now that it has come into the Stone-Platt group, more operating capital has become available. On the other hand, the new ownership of the group has slightly redirected Fluidfire's activities, which now give greater emphasis to immediately commercial lines and less to innovation. Virr and his deputy, Trevor Keirle, view the situation philosophically. They accept that after an initial rush of innovation, it has become appropriate to stress the less heady but more long-term consolidation of earlier work. However, they have no doubt that, despite its relatively small size, Fluidfire will continue to be a significant force in fluidized-bed development.

Address   Stone-Platt Fluidfire Limited
             Netherton, Dudley, West Midlands, England   DY2 95E

Financial   No financial data available.
Data

Appendix

Additional Organizations Involved in
Fluidized-Bed Energy Technology Worldwide

ARGONNE NATIONAL LABORATORY

Program   Argonne National Laboratory is one of the largest federally funded research and development laboratories in the United States, with an annual budget of $150 million. It is operated under an agreement between the Department of Energy (DOE), the University of Chicago, and the Argonne Universities Association (AUA, a group of 30 Midwestern universities). Argonne's facilities are owned by the federal government, but are operated by the University of Chicago for DOE, according to policies established jointly with the AUA.

Argonne has done dozens of support studies on sorbent utilization and regeneration for fluidized-bed combustors. The Laboratory has also completed the conceptual design, and is overseeing the final design and construction work, on a 3 MW pressurized fluidized-bed Component Test and Integration Unit (CTIU). This flexible test rig will be used to support DOE's pressurized fluidized-bed development program and to investigate alternative concepts of pressurized fluidized-bed combustion. Argonne hopes to have the CTIU completed by 1981.

Address   9700 South Cass Avenue, Argonne, Illinois 60439
(312) 972-2000

Contact   Albert A. Jonke

UNIVERSITY OF ASTON IN BIRMINGHAM

Program   The University of Aston, site of much of the seminal fluidized-bed combustion work of the late Professor Douglas Elliott, is presently examining the characteristics of a rotating fluidized-bed combustor. Experiments have been conducted burning both gas and anthracite.

Address   Department of Mechanical Engineering, University of Aston Birmingham, England

Contact   Dr. J. R. Howard

BABCOCK HITACHI COMPANY

Program   The Babcock Hitachi Company has been using a 3.4 square foot test unit to examine the issue of nitric oxide emissions from fluidized-bed combustors. The company feels that, on the basis of the experiments it has done, emissions can be controlled using a two-stage combustion process.

Address   Takaramachi 3-36, Kure R.L., Kure, Hiroshima
           Japan

Contact   Hiroshi Terado

BERGBAU-FORSCHUNG GMBH

Program   Bergbau-Forschung GMBH, a coal research organization, is designing under the sponsorship of the West German Ministry of Research and Technology, an open gas turbine cycle utilizing a pressurized fluidized-bed combustor. Presently, the organization is operating a small test unit to determine parameters for the move to larger scale models.

Address   4300 Essen 13, Frillendorfer Strasse 351
           West Germany

Contact   Dr. H. Schreckenberg

BHARAT HEAVY ELECTRICALS LIMITED

Program   Bharat Heavy Electricals Limited, an Indian electrical engineering manufacturer, has under construction a 6.6 foot diameter atmospheric fluidized-bed combustor. The company sees fluidized-bed systems as an effective means of burning low-grade high-ash Indian coal for electricity generation.

Address   New Delhi, India

Contact   Bhupinder Gill

BROOKHAVEN NATIONAL LABORATORY

Program   Brookhaven National Laboratory is a research and development facility managed by Associated Universities, Inc. (a nonprofit corporation formed by nine northern universities) for DOE. The Laboratory is presently conducting experiments on the regeneration of spent sorbents from fluidized-bed combustors.

Address   Upton, Long Island, New York 11973
            (516) 345-2123

Contact   Ralph T. Young

CENTRAL MECHANICAL ENGINEERING RESEARCH INSTITUTE

Program   The Central Mechanical Engineering Research Institute has several test units which it is using to experiment with the fluidized-bed combustion of Indian coal. Indian coal is typically 40 percent ash and 1 percent sulfur. The Institute is attempting to develop a fluidized-bed package boiler to convert oil-fired systems to coal-firing.

Address   Durgapur, India

Contact   Prabir Basu

CLEAVER BROOKS

Program   Cleaver Brooks, a division of Aqua-Chem, Inc., manufactures packaged boiler systems. It has supplied the Alexandria, Virginia test facility of Pope, Evans and Robbins Incorporated with two fluidized-bed boilers--a 5,000 pound-per-hour unit in 1965, and, recently, a more sophisticated boiler.

Address   P.O. Box 421, Milwaukee, Wisconsin 53201
            (414) 962-0100

Contact   Robert Chronowski

COPELAND SYSTEMS INCORPORATED

Program   Copeland Systems is an engineering concern specializing in the design of fluidized-bed combustion systems. The Fluidized Combustion Corporation owns 51 percent of Copeland Systems and Foster Wheeler (see profile) owns 25 percent. The company has seventeen years of experience with fluidized bed incinerators, and has built sixty commercial plants around the world.

Address   708 Enterprise Drive, Oak Brook, Illinois 60521
(312) 654-2820

Contact   Ian Lutes

CSIRO

Program   CSIRO has two fluidized-bed test rigs, a nine inch diameter combustor and a one-by-one foot unit, which it is using to burn Austrailian coals. Research thus far has concentrated on examining the heat transfer properties of fluidized-bed combustors.

Address   P.O. Box 225, Dickson, ACT 2602, Austrailia

Contact   Sir Robert Price

ELECTRIC POWER RESEARCH INSTITUTE

Program   The Electric Power Research Institute (EPRI) is the research and development organization of the utility industry. The 1978 program funding for EPRI is $193 million, of which $6 million is earmarked for research on fluidized-bed combustion and coal cleaning.

EPRI has funded the construction of a six-by-six foot atmospheric fluidized-bed development unit at the Babcock & Wilcox Company's Research Center in Alliance, Ohio (see Babcock & Wilcox [U.S.] profile). This unit is designed to provide information on adapting atmospheric fluidized-bed combustion technology for utility use. EPRI is also funding a comprehensive corrosion/erosion test program and support studies on various aspects of pressurized fluidized-bed combustion, particularly hot gas clean-up.

Address	3412 Hillview Avenue, P.O. Box 10412, Palo Alto, California 94303 (415) 855-2000
Contact	Shelton Ehrlich

THE ENERGY EQUIPMENT COMPANY LIMITED

Program	Energy Equipment Company Limited is a small boiler manufacturer located in Olney, Buckinghamshire, England. The company has developed an atmospheric fluidized-bed combustion system which it is presently offering commercially in sizes up to 40,000 pounds per hour. One such unit, a 30,000 pound-per-hour boiler, has been installed at a Cadbury Limited plant in Bournville, Birminghamshire, England, and has been operating successfully since March 1978. The company says that several other organizations have indicated interest in purchasing units.
Address	Olney, Buckinghamshire, England
Contact	P.B. Caplin

ENERGY INCORPORATED

Program	Energy Incorporated is a small company which designs and manufactures radioactive waste volume reduction systems, nuclear plant safety mechanisms, and fluidized-bed incinerators for energy recovery. Energy Products of Idaho, a majority-owned subsidiary of Energy Incorporated, has sold almost two dozen of its fluidized-bed incinerator systems, called Fluid Flame, in sizes that produce 10,000 to 30,000 pounds-per-hour of steam. These systems burn wood waste, and research has been done for burning such combustible waste materials as olive pits, corn cobs, almond shells, and date pits in these systems.
Address	P.O. Box 736, Idaho Falls, Idaho 83401 (208) 524-1000
Contact	Eric Pedersen

EXXON CORPORATION

Program	Exxon is a multinational oil company. Its sales in 1977 were $54 billion. The Exxon Research and Development Company, under a contract with the Environmental Protection Agency, has designed, constructed and is now operating a pressurized fluidized-bed test unit, called the Miniplant. Currently, this unit is being used to compile and analyze fluidized-bed combustion-emissions data.

As a part of the Department of Energy demonstration program for industrial applications of fluidized-bed combustion processes, Exxon has a contract to design, build and operate a fluidized-bed indirect-fired process heater for use in an oil refinery. Phase 1 of this contract, which involves preliminary testing and design studies, is presently underway. |
| Address | 1251 Avenue of the Americas, New York, New York 10020 (212) 398-3000 |
| Contact | Rene Bertrand |

GENERAL ELECTRIC COMPANY

Program	Under a $4.38 million Department of Energy contract, General Electric Energy Systems, a division of General Electric Company, is developing a coal-fired combined-cycle power concept. The combined steam- and gas-power cycle utilizes a pressurized fluidized-bed combustor that burns coal. General Electric also has a $3 million DOE contract to do design development of a full-size turbine subsystem under DOE's High Temperature Turbine Technology Program (which examines the critical problems of a large coal-fired, high-temperature, high-efficiency, gas-turbine engine).
Address	River Road, Schenectady, New York 12345 (518) 385-2211
Contact	Robert D. Brooks

GEORGETOWN UNIVERSITY

Program   Georgetown University, a Jesuit-run university in Washington, D.C., has a contract with the Department of Energy as a part of DOE's demonstration program for industrial applications of fluidized-bed combustion systems. The contract was the result of several years of effort by Georgetown to obtain federal assistance in building a coal-fired boiler. The contract calls for the design, construction and operation of a 100,000 pounds-per-hour atmospheric fluidized-bed boiler. Georgetown has subcontracted the work to Fluidized Combustion Company, a subsidiary of Pope, Evans and Robbins Incorporated and Foster Wheeler Corporation. The unit is scheduled to be on line in 1978.

Since the Georgetown area of Washington is an historic district noted for the beauty of its environment, those involved in the Georgetown fluidized-bed project feel that it is an excellent testing ground for burning coal in an urban setting. They say that if they can sell the idea of burning coal in Georgetown, they can sell it anywhere.

Address   37th and O Streets NW, Washington, D.C.   20057
(202) 625-0100

Contact   Ben Scarbrough

GRAND FORKS ENERGY RESEARCH CENTER

Program   Grand Forks Energy Research Center, a DOE laboratory, has been conducting tests in its experimental six-inch fluidized bed combustor on several different lignites and Western subbituminous coals. The tests were performed to evaluate the effects of coal composition on the retention of sulfur in the fluidized bed.

Address   Box 8213, University Station, Grand Forks, North Dakota 58202
(701) 775-4207

Contact   Gerald M. Goblirsch

HELSINKI UNIVERSITY OF TECHNOLOGY

Program The Laboratory of Energy Economics and Power Plants at the Helsinki University of Technology has been experimenting with the fluidized-bed combustion of peat since 1975. Peat, because of its high (50 percent) moisture content, is difficult to burn by conventional means. It does, however, have the advantages of low ash and low sulfur content. The Finnish scientists and engineers have experience working with fluidized-bed technology gained from metallurgical work. The goal of the Helsinki University of Technology fluidized-bed combustion program is to develop a peat-fueled district heating plant.

Address Helsinki, Finland

Contact A. Jahkola

INSTITUTE OF GAS OF THE UKRAINIAN SSR ACADEMY OF SCIENCE

Program The Institute of Gas of the Ukrainian SSR Academy of Science is doing research in the combustion of gas in fluidized beds. Pilot units have combusted a prepared mixture of gas and air, gas partially mixed with air, and gas with air injected into the combustor separately. These tests were run in a 2.2 meter diameter test combustor.

Address Kiev, USSR

Contact K. Ye Makhorin

THE INTERNATIONAL BOILER WORKS COMPANY

Program International Boiler Works, a subsidiary of Combustion Equipment Associates, Inc., is manufacturing a fluidized-bed combustor from an Energy Resources Company (see profile) design.

Address 460 Birch Street, East Stroudsburg, Pennsylvania 18301 (717) 421-5100

Contact Fred Taylor

IOWA STATE UNIVERSITY

Program | Iowa State is analyzing the thermodynamics of processes for regenerating spent limestone sorbent from a fluidized-bed combustion unit.

Address | Ames, Iowa 50011
(515) 294-4111

Contact | Thomas D. Wheelock

JOHNSTON BOILER COMPANY

Program | Johnston Boiler Company, a small, privately owned boiler manufacturer, is the United States licensee for Combustion Systems Limited fluidized-bed combustion units (see Combustion Systems Limited profile). Johnston has installed a prototype 10,000 pound-per-hour fluidized-bed steam generator at its home office in Michigan, and the unit is currently undergoing extensive testing. Johnston hopes to market packaged fluidized-bed boiler systems to industrial customers.

Address | Ferrysburg, Michigan 49409
(616) 842-5050

Contact | R. R. Whitehouse

KANSAS STATE UNIVERSITY

Program | Kansas State University is conducting experiments on the fluidized-bed gasification of agricultural waste material.

Address | Manhattan, Kansas 66506
(913) 532-6011

Contact | Walter P. Walawender

LEEDS AND NORTHRUP COMPANY

Program   Leeds and Northrup Company, a designer and manufacturer of electronic instrumentation and process control systems, has developed under DOE sponsorship an instrument to measure and analyze particulates in the gas stream emerging from a pressurized fluidized-bed combustor.

Address   Dickerson Road, North Wales, Pennsylvania 19454
            (215) 643-2000

Contact   E. C. Muly

LEHIGH UNIVERSITY

Program   Lehigh University undertook experimentation with centrifugal fluidized-bed combustion in 1974. This is a relatively new concept for coal combustion, where the bed rotates about its vertical axis of symmetry and the fluidizing air flows radially inward through the porous cylindrical surface of a distributor. The Lehigh program is presently being sponsored by DOE.

Address   Packard Lab, Building No. 19, Bethlehem, Pennsylvania 18015

Contact   Edward Levy

UNIVERSITY OF MARYLAND

Program   The University of Maryland's Department of Chemical Engineering is conducting experiments in optimizing the utilization of limestone as a sorbent in fluidized-bed combustion units.

Address   Department of Chemical Engineering, College Park, Maryland 20742
            (301) 454-2431

Contact   Larry Gasner

MASSACHUSETTS INSTITUTE OF TECHNOLOGY

Program   Massachusetts Institute of Technology's Department of Chemical Engineering is conducting fluidized-bed modelling studies and experiments on nitrogen oxide emissions.

Address   Room 26-147, Cambridge, Massachusetts 02139
(617) 253-4579

Contact   Shao E. Tung

MITRE CORPORATION

Program   The MITRE Corporation, a federal contract resource center, provides technical assistance services to government agencies.   The METREK division of MITRE has provided services to the Department of Energy for its fluidized-bed combustion development program, particularly the Rivesville project (see Pope, Evans and Robbins Incorporated profile).

Address   1820 Dolly Madison Boulevard, McLean, Virginia 22101
(703) 827-6000

Contact   Charles Bliss

MONONGAHELA POWER COMPANY

Program   A 30 MWe "multi-cell" fluidized-bed combustion unit has been installed at the Rivesville, West Virginia plant of the Monongahela Power Company, a subsidiary of the Allegheny Power System, Inc. Pope, Evans and Robbins Incorporated designed the unit and DOE funded it (see Pope, Evans and Robbins profile).

Address   Fairmont Avenue, Fairmont, West Virginia 26554
(304) 366-3000

Contact   Homer McCarthy

MORGANTOWN ENERGY RESEARCH CENTER

Program
Morgantown Energy Research Center, a DOE research and development laboratory, is the project manager for a 6 MW atmospheric fluidized-bed Component Test and Integration Unit (CTIU). The design calls for three fluidized-bed cells to be stacked one on top of another. The bottom two cells will burn coal; the top, the carbon burn-up cell, will burn the residual carbon in the ash from the first two cells. The CTIU will serve as a flexible test rig to address and resolve key technical issues which have arisen in the DOE fluidized-bed development program. It will also provide design data for future stacked bed applications.

Morgantown has also conducted experiments in the fluidized-bed combustion of Texas lignite. Lignite composes 29 percent of the United States' solid fuel reserves and the Morgantown test results indicate that lignite may be a particularly attractive fuel for combustion in a fluidized-bed unit. DOE is considering a lignite demonstration project.

Address
Culensberry Road, Morgantown, West Virginia 26505
(304) 599-7764

Contact
Augustine A. Pitrolo

MUSTAD AND SØN AS

Program
A Norwegian company, Mustad and Søn AS, has been involved in fluidized-bed combustion research and development since it collaborated in the 1960s with the Norwegian Institute of Technology on a fluidized-bed incinerator for hospital waste. Mustad and Søn designed and built the combustor now in use at the AB Enkoepings Vaermeverk district heating installation in Sweden (see AB Enkoepings Vaermeverk profile) and is offering industrial-size fluidized-bed units commercially.

Address
Mustad Rd 1, Oslo, Norway

Contact
H. Eckhoff

NAGOYA UNIVERSITY AND NAGOYA INSTITUTE OF TECHNOLOGY

Program   Nagoya University and Nagoya Institute of Technology have developed mathematical models of the fluidized-bed combustion of coal which factor in the size distribution of fuel feed, the moisture, ash and sulfur content of the fuel, elutriation, ash off-take, and partial combustion. The key question for Japanese fluidized-bed units is how to achieve low nitrogen oxide emissions, since this is a particular problem in Japan.

Address   Department of Iron and Steel, Nagoya University
Ferro-Cho, Chikusa-Ku, Nagoya, Japan 464

Contact   Masayuki Horio

NATIONAL AERONAUTICS AND SPACE ADMINISTRATION

Program   The Lewis Research Center of the National Aeronautics and Space Administration (NASA) managed the Energy Conversion Alternatives Study (ECAS) for the Energy Research and Development Administration (ERDA, now the Department of Energy) and the National Science Foundation. The overall objective of the project was to study advanced power-generation techniques which could use coal or coal-derived fuels and to evaluate the relative merits and potential benefits of such techniques. The total program funding was $2.6 million; ERDA contributed $1.6 million and the National Science Foundation $1 million. The contractors were the General Electric Company, Westinghouse Electric Corporation, a team of Burns and Roe, Inc. and United Technologies, Inc. The program was finished in 1977.

The Lewis Research Center also has a pressurized fluidized-bed test rig, which is used for technical support studies. It will provide data for a computer model of the fluidized-bed combustion process which will be used to extrapolate the NASA test results to larger-scale fluidized-bed units.

Address   2100 Brook Park, Cleveland, Ohio 44134
(216) 433-4000

Contact   Richard J. Priam

NEW YORK UNIVERSITY

Program	The Division of Aerospace and Energetics of New York University is using its one-foot diameter test fluidized-bed combustor to investigate the effects of different types of tubing on heat transfer coefficients in fluidized-bed units. The work is being funded by DOE.
Address	Department of Applied Sciences, Division of Aerospace and Energetics, New York University, 26-36 Stuyvesant Street, Westbury, Long Island, New York 10003 (516) 832-2599
Contact	Victor Zakkay

OREGON STATE UNIVERSITY

Program	Oregon State University is using its one meter square fluidized-bed test unit to investigate the effects of boiler tube spacing, boiler tube location, bank location, and fluidizing gas velocity on solids mixing in a fluidized bed.
Address	Corvallis, Oregon 97331 (503) 754-0123
Contact	Thomas Fitzgerald

RADIAN CORPORATION

Program	DOE is sponsoring a program to develop conceptual designs for a 600 MWe utility atmospheric fluidized-bed combustion steam-generating plant in order to obtain design information for a demonstration unit. Conceptual designs are being developed by the following teams: Burns and Roe/Combustion Engineering; Stone & Webster/Foster Wheeler; and Stone & Webster/Babcock & Wilcox (U.S.). Radian Corporation, a technical consulting firm, is serving as a support contractor to DOE on the project. Radian is providing technical assistance of periodic reviews of progress on the three designs.
Address	Suite 125, Grant Building, 1651 Old Meadows Road, McLean, Virginia 22101 (703) 821-8866

Contact   D. N. Garner

RALPH STONE AND COMPANY, INC.

Program   Ralph Stone and Company, Inc., under an Environmental
               Protection Agency contract, is conducting an environmental
               assessment of the residues resulting from both the fluid-
               ized-bed combustion of coal and the gasification of high-
               sulfur fuel oils.

Address   10954 Santa Monica Boulevard, Los Angeles, California
               90025
               (213) 478-1501

Contact   Ralph Stone

REYNOLDS, SMITH AND HILLS

Program   Reynolds, Smith and Hills is the architect-engineer, and
               will direct the construction work, for the atmospheric
               fluidized-bed Component Test and Integration Unit being
               built at DOE's Morgantown Energy Research Center (see
               Morgantown Energy Research Center in this appendix).

Address   4019 Boulevard Center Drive, Jacksonville, Florida 32207
               (904) 396-2011

Contact   C. C. Space

SPECTRON DEVELOPMENT LABORATORIES, INC.

Program   Spectron Laboratories, under DOE sponsorship, has devel-
               oped particulate diagnostics instrumentation for use in
               fluidized-bed combustion plants. Particle field measure-
               ments are needed to determine efficiencies of particle sepa-
               rators and filters, and to measure particles at inlets to gas
               turbines.

Address   3303 Harbor Boulevard, Costa Mesa, California 92626
               (714) 549-8477

Contact   William D. Bachalo

STERNS-ROGER INC.

Program   Under the technical direction of Argonne National Labora-
           tory, Sterns-Roger Inc. is preparing the preliminary design
           for a pressurized fluidized-bed combustion Component Test
           and Integration Unit to be built at Argonne (see Argonne
           National Laboratory in this appendix).

Address   700 South Ash Street, Denver, Colorado 80217
           (303) 770-6400

Contact   Donald K. Clark

STONE & WEBSTER, INC.

Program   Stone & Webster, Inc., a large engineering and construction
           firm, has negotiated an agreement with Pope, Evans and
           Robbins Incorporated to provide joint marketing of consult-
           ing engineering services for fluidized-bed boiler plants.
           Stone & Webster is now working on two Department of Energy
           contracts for 570-MWe atmospheric fluidized-bed combustion
           unit conceptual designs. Stone & Webster has teamed with
           Foster Wheeler for one design and the Babcock & Wilcox
           Company (U.S.) for the other (see Foster Wheeler and
           Babcock & Wilcox [U.S.] profiles).

Address   90 Broad Street, New York, New York 10004
           (212) 269-4224

Contact   T.G. Wells

TENNESSEE VALLEY AUTHORITY

Program   The Tennessee Valley Authority, a government-operated
           utility, has contracted with three boiler manufacturers--
           Combustion Engineering, Inc., Babcock & Wilcox Company
           (U.S.), and Foster Wheeler Corporation (see profiles)--for
           a design of a 200 MWe atmospheric fluidized-bed demonstra-
           tion plant.

Address   Chattanooga, Tennessee 37401
           (615) 755-3011

Contact   Harold Falkenberry

U.S. DEPARTMENT OF AGRICULTURE

Program   The Department of Agriculture is evaluating the solid waste from fluidized-bed combustion processes for potential use in agriculture. It is making detailed chemical and physical analyses of the material, followed by extensive greenhouse, growth-chamber, and field studies.

Address   Agricultural Research Service, West Virginia University
Morgantown, West Virginia
(304) 599-7186

Contact   Dr. O. L. Bennett

U.S. ENVIRONMENTAL PROTECTION AGENCY

Program   The Environmental Protection Agency (EPA) has governmental responsibility both for assessing the environmental implications of fluidized-bed combustion systems and for developing pollution-control technology for the systems. The EPA program is designed to integrate the appropriate pollution-control measures into DOE's fluidized-bed combustion development program.

EPA is spending $4 million a year on a program that consists of eleven projects carried out by a variety of contractors. Battelle Memorial Institute and TRW Defense and Space Systems Group are working on environmental assessment projects. Projects for the development of pollution-control technologies are split into two areas: 1) experimental and engineering studies, and 2) solid and liquid waste disposal studies. Westinghouse Research and Development Center, Argonne National Laboratory, Exxon Research and Engineering Company, the Aerotherm Division of Acurex Corporation, and the Massachusetts Institute of Technology are carrying out the experimental and engineering studies while Westinghouse, Ralph Stone and Company, and the Tennessee Valley Authority are doing the solid and liquid waste disposal work.

Address   Office of Research and Development, Industrial Environmental Research Laboratory, Research Triangle Park,
North Carolina 27711
(919) 541-2821

Contact   D. Bruce Henschel

VALLEY FORGE LABORATORIES, INC.

Program	Valley Forge Laboratories, a private research laboratory, is conducting studies on potential uses for the residue from the fluidized-bed combustion process. The work is funded by DOE.
Address	6 Berkeley Road, Devon, Pennsylvania 19333 (215) 688-8517
Contact	Richard H. Miller, Sr.

VIRGINIA POLYTECHNIC INSTITUTE AND STATE UNIVERSITY

Program	The coal workshop at Virginia Polytechnic Institute and State University (VPI & SU), directed by Professor Arthur Squires, is carrying out a research program on a 1 million BTU per hour fluidized-bed coal-fired boiler. The boiler was constructed for VPI & SU by Stone-Platt Fluidfire Limited (see Stone-Platt Fluidfire profile).
Address	Department of Chemical Engineering, Blacksburg, Virginia 24061 (703) 951-5972
Contact	Professor Arthur Squires

WEST VIRGINIA UNIVERSITY

Program	West Virginia University's (WVU) Department of Chemical Engineering has under construction a two-by-two foot fluidized-bed combustion test rig. WVU also has a two-by-two cold model test unit which is being used to study particle mixing in a fluidized bed.
Address	Morgantown, West Virginia 26506 (304) 293-0111
Contact	C. Y. Wen

WESTINGHOUSE ELECTRIC CORPORATION

Program
Westinghouse Electric Corporation manufactures and sells equipment for the generation, transmission, utilization and control of electricity. Since 1972, Westinghouse Research and Development Center in Waltz Mills, Pennsylvania has been developing a fluidized-bed process to convert coal to a low-BTU gas capable of being used either as an industrial fuel or as a fuel to power a combined-cycle electric generating plant. The project is funded by DOE.

Westinghouse is also involved in a number of support studies investigating environmental aspects of fluidized-bed combustion. These include an assessment of the impact of more stringent sulfur dioxide emission standards on the economics of fluidized-bed combustion units; an evaluation of the effect sorbent regeneration would have on the economics of fluidized-bed combustion; an assessment of alternative sorbents; an analysis of particulate-control options, and an evaluation of the trace-element problems which fluidized-bed systems might be facing.

Address
Gateway Center, Pittsburgh, Pennsylvania 15222
(412) 255-3800

Contact
Dale L. Keairns

UNIVERSITY OF WISCONSIN--MILWAUKEE

Program
The University of Wisconsin--Milwaukee is analyzing a new particle-collection technique for hot-gas clean-up systems for pressurized fluidized-bed combustion units.

Address
3203 North Downer Street, Milwaukee, Wisconsin 53201
(414) 963-1122

Contact
Keh C. Tsao

WORMSER ENGINEERING, INC.

Program	Wormser Engineering is a small, four-year-old firm which was formed specifically to develop fluidized-bed units for industrial applications. Wormser presently has a demonstration fluidized-bed unit undergoing shakedown in a Lynn, Massachusetts factory, where it is being used for space heating. Wormser hopes to offer a line of fluidized-bed boilers in the three to ten MW (thermal) range. Wormser feels its design will be able to meet all air pollution requirements.
Address	50 High Street, Lynn, Massachusetts 01902 (617) 581-7580
Contact	Alex Wormser

YORK-SHIPLEY, INCORPORATED

Program	York-Shipley, Incorporated, a subsidiary of CompuDyne Corporation, is a manufacturer of steam generators, furnaces, and rooftop combination heating and air conditioning units. York-Shipley is the Eastern United States licensee for Energy Incorporated's fluidized-bed combustion systems (see Energy Incorporated in this appendix).
Address	693 North Hills Road, York, Pennsylvania 17405 (717) 755-1081
Contact	C. H. Neiman, Jr.

Methodology

Background research for <u>Fluidized-Bed Energy Technologies: Coming To A Boil</u> began on a part-time basis in June, 1977, and in earnest in September. An intensive survey of both United States and European fluidized-bed literature was done. Among the most useful sources were the proceedings of several conferences: the International Conferences on Fluidized-Bed Combustion, the Fluidized-Bed Combustion Technology Exchange Workshop, and the Workshop on Utility/Industrial Implementation of Fluidized-Bed Combustion Systems. The fluidized-bed publications of the U. S. Energy Research and Development Administration (particularly the quarterly <u>Power & Combustion Report</u>), the Electric Power Research Institute, and the U. S. Environmental Protection Agency were also key. Useful journals and trade publications included <u>Science</u>, <u>Chemical Engineering</u>, <u>Business Week</u>, and <u>Coal Age</u>.

Between June and October, 1977, INFORM identified from these sources the organizations heavily involved in fluidized-bed research. These organizations were contacted by letter and asked for an annual report and any press releases and technical papers available on their fluidized-bed combustion program. The top fluidized-bed personnel at some groups were then contaced by phone and meetings were arranged for late November, when author Walter Patterson, INFORM's London-based researcher on the project, arrived for an intensive two-week period of interviews.

The interviews were based upon a questionnaire (reproduced at the end of this section) which was designed to ferret out the history of the organization's involvement in the technology, its present program, its goals for

119

the future, and its thoughts on the economic and institutional factors affecting commercial prospects of fluidized-bed energy technologies. The organizations who were not interviewed were contacted by phone and asked to provide the same information.

Almost all of the organizations contacted by INFORM responded enthusiastically to the chance to describe their fluidized-bed program. They were, for the most part, very generous with their time and often provided additional information beyond that which was requested.

In December, Walter Patterson returned to the United Kingdom where he conducted face-to-face and telephone interviews with European fluidized-bed teams. Co-author Richard Griffin, INFORM's New York-based researcher on the project, attended the Fifth International Conference on Fluidized-Bed Combustion in Washington, D.C. on December 12-14.

The profiles were written during the winter using interview notes, technical and public-relations material supplied by the companies, and public sources. INFORM submitted the profiles to the interviewees for review of factual accuracy and update of recent occurrences. INFORM used the responses, usually written but occasionally given over the phone, to update the profiles, and it edited the comments if given reasonable justifications for so doing.

The introductory material was written in the late fall and winter. Patterson returned to the United States at the end of March for several days of rewrite and the introductory material was sent out for review in April to fluidized-bed experts at the U.S. Department of Energy, the U.S. Environmental Protection Agency, and the Electric Power Research Institute and to several knowledgeable individuals from the corporate, academic and public-interest sectors. These reviewers' comments were incorporated into the text, and final production completed in May, 1978.

STUDY OF FLUIDIZED-BED ENERGY TECHNOLOGY

Questionnaire

Name and address of organization:

1. Why did your organization become involved with this technology?

2. Could you provide a brief history of your research and development projects or commercial undertakings to date?

    a) Which part or parts of your organization is/are doing the work?

    b) What milestones have been achieved and when?

    c) How many scientists and engineers are involved on a full-time basis? Are there other staff members--economists, marketing or other commercial personnel--similarly involved?

    d) How is the work funded? (If the funds come from more than one source please specify approximate percentage breakdown.)

    e) How much has your organization spent on research and development to date?

3. What is the current status of your work?

    a) What are the current technical goals?

    b) What is the timetable for future research?

    c) Have proposals for R&D or commercial contracts been submitted? (If so, to whom?)

    d) Have contracts been signed for forthcoming work? (If so, with whom, and what are the terms?)

e) Are you designing, constructing or operating one or more fluidized-bed units at present? If so, please give technical specifications and other details as listed (to be completed for each separate unit or system).

1. Name of unit or system

2. Location

3. Organization or organizations involved

4. Purpose of unit or system

5. Physical description

    a) area of bed   b) bed depth in operation   c) bed material   d) velocity of fluidization   e) operating pressure   f) bed additives if any   g) fuel feed system   h) heat removal system   i) operating temperature range   j) heat output in operation

6. Date of first operation

7. Specifications of ancillary systems if applicable (turbine etc.)

8. Operating record to date

9. Fuel characteristics

10. Emission characteristics (sulfur and nitrogen oxides, particulates, alkali metals, vanadium and other trace elements)

11. Current research/development/commercial objectives

12. Technical problems as yet unresolved

13. Economic prognosis: capital costs, running costs, market prospects and status

4. What is the goal of your program? Do you view it as short-term, medium-term, or long-term?

5. What factors are inhibiting your progress (technical, economic, environmental, institutional)? If possible please rank them in order of importance.

6. What factors are encouraging your progress (technical, economic, environmental, institutional)? If possible please rank them in order of importance.

7. How do you view your role in the development of this technology?

8. How do you view the role of other organizations (i.e., the government) in the development of this technology?

9. What role do you see for fluidized-bed combustion in the future (5, 10, 20 years hence) within the U.S. national energy scene? internationally?

10. Who in the organization makes the decisions regarding the program, and how?

11. What influences are taken into account--for instance business prospects, R&D funding assistance, investment implications, environmental implications, the role of the technology in U.S. energy policy, etc.?

12. What costing has been done, and how? Are costings still prospective or already contractual? Has consideration been given to capital cost of plant, fuel cost, cost for environmental measures (for instance sorbent supply and disposal), maintenance, comparison with other technological options (traditional boilers, boilers with scrubbers, nuclear plant, etc.)?

13. Are patent considerations a problem? What is the patent status of work being done by the organization? Are patents held, or applied for? What about licensing? Do conflicts arise because of government R&D funding, or between industrial organizations?

14. What is the position on warranties? Is the reliability of existing designs demonstrated? Do current contracts offer warranty? On what basis?

15. What are the implications of scale-up from existing designs? Will modular construction and/or replication have advantages?

16. What are the prospects for export of the technology? Have export credit agencies shown any interest?

Bibliography

Books and Reports

American Institute of Physics. Efficient Uses of Energy. New York: U.S. Energy Research and Development Administration, 1975.

Biniek, Joseph P., and Gulick, Frances A. The National Energy Plan: Options Under Assumptions of National Security Threat OR Energy Policy As If It Really Mattered. Washington, D.C.: U.S. Government Printing Office, 1978.

Cannon, James S., and Herman, Stewart W. Energy Futures: Industry and the New Technologies. New York: INFORM, Inc., 1976.

Cart, E.N.; Farmer, M.H.; Jahnig, C.E.; Lieberman, M.; and Spooner, F.M. Evaluation of the Feasibility for Widespread Introduction of Coal into the Residential and Commercial Sectors. Linden, N.J.: Exxon Research and Engineering Company, 1977.

Electric Power Research Institute. Comparative Study and Evaluation of Advanced Cycle Systems. Palo Alto, Cal.: EPRI, 1976.

Electric Power Research Institute. Program Structure and Funding: 1978. Palo Alto, Cal.: EPRI, 1978.

Energy Resources Company. Effects of Setting New Source Performance Standards For Fluidized-Bed Systems. Washington, D.C.: U.S. Environmental Protection Agency, 1978.

Environmental Information Center, Inc. Energy Directory Update. New York: Environmental Information Center, Inc., 1977.

Executive Office of the President, Energy Policy and Planning. The National Energy Plan. Washington, D.C.: U.S. Government Printing Office, 1977.

Farmer, M.H.; Magee, E.M.; and Spooner, F.M. Applications of Fluidized-Bed Technology to Industrial Boilers. Linden, N.J.: Exxon Research and Engineering Company, 1977.

126

Fowler, John M. Energy-Environment Source Book. Volumes 1-2. Washington, D.C.: National Science Teachers Association, 1977.

Fuel and Energy Consultants, Inc. Survey and Analysis of Fluidized-Bed Combustion Work Being Conducted in Europe. Washington, D.C.: U.S. Energy Research and Development Administration, 1976.

The General Electric Company. Energy Conversion Alternatives Study: General Electric Phase 2 Final Report. Volumes 1, 2, and 3. Washington, D.C.: National Aeronautics and Space Administration, 1976.

Grant, Andrew J., and LaNauze, Robert D. "The Impact of Fluidized-Bed Combustion on United States Coal Mining and Use." Paper given at the National Coal Association/Bituminous Coal Research Coal Conference and Expo IV, Louisville, Ken., 1977.

Hoke, R.C.; Bertrand, R.R.; Nutkis, M.S.; Kinzler, D.D.; Ruth, L.A.; Iccarino, E.P.; and Gregory, M.W. Studies of the Pressurized Fluidized-Bed Coal Combustion Process. Washington, D.C.: U.S. Environmental Protection Agency, 1976.

Jonke, Albert A. "Fluidized-Bed Combustion: A Status Report" in Coal Processing Technology. Volume 2. New York: Chemical Engineering Progress, 1975.

Lovins, Amory B. Soft Energy Paths: Toward a Durable Peace. Cambridge, Mass.: Ballinger Publishing Company, 1977.

Practicing Law Institute. Coal Conversion: Practical and Legal Implications. New York: Practicing Law Institute, 1977.

Proceedings of the Fluidized-Bed Combustion Technology Exchange Workshop. Volumes 1 and 2. McLean, Va.: The MITRE Corporation, 1977.

Proceedings of the Fourth International Conference on Fluidized-Bed Combustion. McLean, Va.: The MITRE Corporation, 1976.

Proceedings of the Workshop on Utility/Industrial Implementation of Fluidized-Bed Combustion Systems. McLean, Va.: The MITRE Corporation, 1976.

Program for the Fifth International Conference on Fluidized-Bed Combustion. McLean, Va.: The MITRE Corporation, 1977.

Reese, Alexander. Industrial Energy Conservation: Where Do We Go From Here? New York: INFORM, Inc., 1977.

Squires, Arthur M. Development Opportunities for Small- and Medium-Scale Technologies for Utilization of Coal. Washington, D.C.: Council on Environmental Quality, 1976.

Squires, Arthur M. "Expanding the Use of Coal." Paper given at City College of New York, 1978.

U.S. Department of Energy. Coal: Power and Combustion -- Quarterly Report, January-March 1977 . Washington, D.C.: DOE, 1977.

U.S. Energy Research and Development Administration. Background Information on Implementation of Fluidized-Bed Combustion Systems. McLean, Va.: The MITRE Corporation, 1976.

U.S. Energy Research and Development Administration. Coal: Conversion and Utilization -- 1976 Technical Report . Washington, D.C.: U.S. ERDA, 1977.

U.S. Energy Research and Development Administration. Coal: Power and Combustion -- Quarterly Report, July-September 1976. Washington, D.C.: U.S. ERDA, 1977.

U.S. Energy Research and Development Administration. Fossil Energy Research Program of the Energy Research and Development Administration -- FY 1977. Washington, D.C.: U.S. Government Printing Office, 1976.

U.S. Energy Research and Development Administration. Fossil Energy Research Program of the Energy Research and Development Administration -- FY 1978. Washington, D.C.: U.S. Government Printing Office, 1977.

U.S. Energy Research and Development Administration. Fossil Energy Program Report, 1975-1976. Washington, D.C.: U.S. ERDA, 1976.

U.S. Energy Research and Development Administration. Market Oriented Program Planning Study (MOPPS) Review Draft. Washington, D.C.: U.S. ERDA, 1977.

U.S. Environmental Protection Agency. Energy/Environment II. Washington, D.C.: U.S. EPA, 1977.

Magazines and Journals

"A boiler that cleans up burning coal." Business Week, 18 August 1975, pp. 88B-88D.

Balzhizer, Richard. "Energy Options to the Year 2000." Chemical Engineering, 3 January 1977.

Burke, Donald P. "FBC may be a better way to burn coal." Chemical Week, 22 September 1976.

Electric Power Research Institute. "Clean Coal: What Does It Cost At the Busbar?" EPRI Journal, November 1976.

Electric Power Research Institute. "Coal: Keystone of Energy Facts." EPRI Journal, August 1977.

"Fluid-bed combustion boilers ordered but generators face severe problems." Energy Research Reports, 15 August 1977, Volume 3, Number 15.

"Fluidized-bed combustion -- full steam !" Environmental Science & Technology, February 1976, Volume 10, p. 120.

Hammond, Allen L. "Coal Research (IV): Direct Combustion Lags Its Potential." Science, 8 October 1976, Volume 194, Number 4261, pp. 172-174.

"Making Coal Burn More Cleanly." Business Week, 10 October 1977, p. 48P.

Squires, Arthur M. "Clean Fuels From Coal Gasification." Science, 19 April 1974, Volume 184, Number 4134, pp. 340-346.

Squires, Arthur M. "Clean Power From Coal." Science, 28 August 1970, Volume 169, Number 3948, pp. 821-828.

Ungeheurer, Friedhel. "Coal's Mr. Clean." Quest, September 1977, pp. 25-29.

U.S. Environmental Protection Agency. FBC Environmental Review: Research in Fluidized-Bed Combustion, Volume 1, Number 1.

Glossary

anthracite   A hard coal of high luster, differing from bituminous in that it contains little volatile matter.

bench-scale   A small-scale laboratory unit for testing process concepts and operating parameters as the first step in the evaluation of a process.

bituminous   A soft coal which, when burned, releases large quantities of volatile matter.

breeder reactor   A nuclear reactor so designed that it produces more fuel than it uses.

BTU   British thermal unit, the quantity of energy required to raise the temperature of one pound of water one degree Fahrenheit.

carbon burn-up cell   An additional, specially-designed boiler region to burn escaped fuel particles. This is designed to improve the combustion efficiency of a fluidized-bed system.

cogeneration   The generation of electricity with direct use of the residual heat for industrial process heat or for space heating.

coke   A residue consisting of carbon and mineral ash formed when bituminous coal is heated in a limited air supply or in the absence of air.

coke breeze   The fine particles left after coke has been crushed.

combined cycle   Two sequential power conversion systems operating at different temperatures.

corrosion   The gradual wearing away of a surface, usually by chemical action.

cyclone collector   An air-pollution control device that uses mechanical means to collect particulates. Also known as a mechanical collector.

cyclone separator	A settling chamber to separate solid particles from a gas.
dolomite	A mineral consisting of calcium magnesium carbonate which is used as a sulfur-trapping agent in some fluidized-bed systems.
electrostatic precipitator	An air-pollution control device that uses an electric field to trap up to 99.5 percent of particulates in a gas stream.
elutriation	The lifting out of the small elements in a mixture of solid particles by a stream of high-speed gas.
fluidization	To suspend particles in a rapidly moving stream of gas or vapor; the particles are close enough together and interact in such a manner that they give the impression of a boiling liquid.
fluidized bed	A gathering of small solid particles maintained in balanced suspension against gravity by the upward motion of a gas.
fluidized-bed combustion	The burning of coal (or some other fuel) in a fluidized bed. The burning may take place at atmospheric pressure, in which case the system is called an atmospheric fluidized-bed combustion unit, or it may take place under several atmospheres pressure, in which case it is called pressurized fluidized-bed combustion.
fluidizing velocity	The rate of upward motion of a gas which is needed to maintain particles in balanced suspension.
gas turbine	A device used in most forms of electric-power generation, composed of blades attached to a central, rotating spindle; the turbine blades are spun by gas at high pressure.
heat exchanger	A device, usually composed of many pipes, in which the heat of one fluid or substance is transferred to another.
hematite	An inert and dense mineral.

lignite   An usually brownish black coal intermediate in hardness between peat and bituminous coal.

limestone   A rock that is formed chiefly by accumulation of organic remains and consists mainly of calcium carbonate. Limestone is used as a sulfur-trapping agent in some fluidized-bed combustion systems.

MW   Megawatt, 1,000,000 watts. A watt is a unit of power. MWe is used as a measurement of output from an electric-generation system, while MW (thermal) is used as a measurement of output from a steam-generation system.

peat   Partially carbonized vegetable tissue formed by partial decomposition in water of various plants.

residence time   The period of time spent by a typical particle in a particular zone of a fluidized bed.

scrubber   A device to remove impurities, especially from gases, before emission to the atmosphere.

sorbent   Particles of crushed limestone or dolomite which trap sulfur by means of a chemical reaction in the fluidized bed.

turn-down ratio   The minimum ratio of actual flowrate to design flowrate at which a fluidized-bed unit can be operated.

Index of Organizations in

Profiles and Appendix

INFORM

Staff

Executive Director
Joanna Underwood

Editorial Director
Jean Halloran

Director of Public Education
Ellen Freudenheim

Administrator
Virginia Jones

Director of Development
Fay Michener

Research Associates
Gino Crocetti
Richard Duffy
Manuel Gomez
Patricia Simko
Vincent Trivelli

Research Assistants
Richard Griffin
Daniel Wiener

Research Consultants
Leslie Allan
Sibyl McC. Groff
Walt Patterson

Editor Frank Stella

Institutional Relations
Joan Baker
Gregory Cohen

Administrative Assistants
Mary Maud Ferguson
Joan Pearson
Amy Pratt
Grace Rubin

Student Intern
Susan Jakoplic

Steering Committee*

Robert Alexander, Consultant
McKinsey & Co.

Marshall Beil, Lawyer
Karpatkin, Pollet & LeMoult

Susan Butler, Administrator
Booth Ferris Foundation

Albert Butzel, Partner
Butzel & Kass

Richard Duffy
INFORM Staff Representative

C. Howard Hardesty, Jr.
Former Vice Chairman, CONOCO

Timothy Hogen, Co-Director
Andover Bicentennial Campaign,
Phillips Academy

Raymond I. Maurice, Ph.D.
Vice President, Tosco Foundation

Dennis Meadows, Director
Research Program, Technology &
Public Policy, Thayer School of
Engineering, Dartmouth College

Barbara Niles, Director
Consumer Action Now

Edward H. Tuck, Partner
Shearman & Sterling

Joanna Underwood
Executive Director, INFORM

Anthony Wolff
Environmental Writer

*affiliations for identification purposes only

Other INFORM Publications

Energy Futures: Industry and the New Technologies
(Ballinger, 660 pages [hardcover], $25.00; abridged edition, 300 pages
[softcover], $8.95)

A survey of corporate research and development of 17 new energy sources
and technologies, including solar, wind, geothermal, trash-to-energy,
coal gasification, and nuclear systems. Environmental impact is evalu-
ated, federal programs assessed, and corporate investment analyzed for
each technology. Over 200 major R&D projects of 142 firms are individually profiled, including technical achievements, and obstacles to commercialization.

Industrial Energy Conservation: Where Do We Go From Here?
(46 pages, $10)

A report on the federal programs and industry progress toward achieving
federal energy conservation targets.

Promised Lands
Volume 1: Subdivisions in Deserts and Mountains (560 pages, $20)
Volume 2: Subdivisions in Florida's Wetlands (540 pages, $20)
Volume 3: Subdivisions and the Law (535 pages, $20)

An analysis of the environmental and consumer impact of the land sales and
subdivision industry. The study examines practices at nineteen large-scale
"new communities" located in the states of Arizona, California, Colorado,
Florida and New Mexico. Laws in these states and on the federal level are
analyzed for how effectively they regulate abuses, such as consumer fraud
and degradation of water resources. Guidelines for upgrading subdivider's
practices are proposed.

Business and Preservation: A Survey of Business Conservation of Buildings and Neighborhoods (300 pages, hardcover, $22; paperback, $14)

An illustrated survey of 71 projects undertaken by American business to
restore and preserve buildings and neighborhoods. Projects analyzed
include adaptive reuse and rehabilitation of historical buildings for new
purposes, neighborhood conservation efforts, and general preservation
support.

To obtain any of these publications, or information on a subscription to
all INFORM publications including its newsletter, contact INFORM,
25 Broad Street, New York, N.Y. 10004, 212/425-3550.

About the Authors

Walter C. Patterson

Walt Patterson is a 40 year-old Canadian, who has been a resident of Britain since 1960. He is the author of Nuclear Power, a Pelican Original published in 1976 by Penguin Books, now in its fourth printing. It was selected as a set book for Britain's Open University and widely and favorably reviewed in Britain and many other parts of the world. He has more recently published The Fissile Society, a study of the social, economic, and political nature of a society in which the major form of energy supply is nuclear electricity. Since 1972, Patterson has been on the staff of the British wing of Friends of the Earth, as energy specialist, in which capacity he has testified before a number of official bodies, including the Parliamentary Select Committee on Science and Technology, the Royal Commission on Environmental Pollution, and the Secretary of State for Energy's National Energy Conference. He is a regular contributor to New Scientist, Energy Policy, Environment, and many other periodicals. He is also a frequent lecturer and broadcaster on television and radio, and a consultant to Parliamentarians and to print and broadcast media.

Richard Griffin

Dick Griffin is a 1977 graduate of Yale University. Hailing from Buffalo, New York, he worked at INFORM during the summer of 1976 and was a member of the 1977 Yale Daily News business board. In his spare time, Griffin has aided the Amalgamated Clothing and Textile Workers Union in their boycott of the J.P. Stevens Company. As a research assistant in energy for INFORM, he has had several speaking engagements including a panel discussion on fluidized-bed coal combustion in Albany, New York. Griffin is presently a candidate for a law degree.